Contents

Addition

Week 1   Using Doubles and Near Doubles 2

Week 2   Using Nice Numbers 12

Week 3   Using Grouping 22

Week 4   Using Partial Sums 32

Week 1   Practice 42

Week 2   Practice 43

Week 3   Practice 44

Week 4   Practice 45

Week 1 — Using Doubles and Near Doubles

Lesson 1

Key Idea

2 + 2 = 4   **Doubles Fact**

2 + 3 = 2 + (2 + 1)   **Near-Doubles Fact**

= 4 + 1 = 5

Try This

Find the sum of each doubles fact.

① 2 + 2 = ____

② 5 + 5 = ____

③ 4 + 4 = ____

④ 1 + 1 = ____

Use a doubles fact to find the sum of each near-doubles fact.

⑤ 2 + 3 = 2 + (2 + 1)
____ + 1 = ____

⑥ 5 + 6 = 5 + (5 + 1)
____ + 1 = ____

⑦ 3 + 4 = 3 + (3 + 1)
____ + 1 = ____

⑧ 4 + 5 = 4 + (4 + 1)
____ + 1 = ____

Author
Sharon Griffin
*Associate Professor of Education and
Adjunct Associate Professor of Psychology*
Clark University
Worcester, Massachusetts

Building Blocks Authors

Douglas H. Clements
*Professor of Early Childhood
and Mathematics Education*
University at Buffalo
State University of New York, New York

Julie Sarama
Associate Professor of Mathematics Education
University at Buffalo
State University of New York, New York

Contributing Writers
Sherry Booth, *Math Curriculum Developer,* Raleigh, North Carolina
Elizabeth Jimenez, *English Language Learner Consultant,* Pomona, California

Program Reviewers

Jean Delwiche
Almaden Country School
San Jose, California

Cheryl Glorioso
Santa Ana Unified School District
Santa Ana, California

Sharon LaPoint
School District of Indian River County
Vero Beach, Florida

Leigh Lidrbauch
Pasadena Independent School District
Pasadena, Texas

Dave Maresh
Morongo Unified School District
Yucca Valley, California

Mary Mayberry
Mon Valley Education Consortium, AIU 3
Clairton, Pennsylvania

Lauren Parente
Mountain Lakes School District
Mountain Lakes, New Jersey

Juan Regalado
Houston Independent School District
Houston, Texas

M. Kate Thiry
Dublin City School District
Dublin, Ohio

Susan C. Vohrer
Baltimore County Public Schools
Baltimore, Maryland

SRAonline.com

Copyright © 2007 SRA/McGraw-Hill.

All rights reserved. Except as permitted under the United States Copyright Act, no part of this publication may be reproduced or distributed in any form or by any means, or stored in a database or retrieval system, without the prior written permission of the publisher, unless otherwise indicated.

Printed in the United States of America.

Send all inquiries to:
SRA/McGraw-Hill
4400 Easton Commons
Columbus, OH 43219-6188

R5313X.01

9 WCE 12 11 10

Photo Credits

3-39 ©PhotoDisc/Getty Images, Inc.

SRA Number Worlds
Addition

Unit 3 Workbook
Level E

featuring Building Blocks Software

Practice

Use the sum of each doubles fact to find the sum of each near-doubles fact.

9. If 3 + 3 = _____, then

3 + 4 = _____.

10. If 5 + 5 = _____, then

5 + 6 = _____.

11. If 7 + 7 = _____, then

7 + 8 = _____.

12. If 8 + 8 = _____, then

8 + 9 = _____.

13. If 5 + 5 = _____, then

5 + 4 = _____.

14. If 2 + 2 = _____, then

1 + 2 = _____.

15. If 7 + 7 = _____, then

6 + 7 = _____.

16. If 4 + 4 = _____, then

3 + 4 = _____.

Find each sum.

17. 3 + 4 = _____

18. 5 + 6 = _____

19. 10 + 9 = _____

20. 7 + 6 = _____

21. 8 + 7 = _____

22. 5 + 4 = _____

Reflect

Given the problem 4 + 5, show how each of the doubles facts below can be used to find the sum.

    4 + 4   or   5 + 5

_____ or _____

Using Doubles and Near Doubles • Lesson 1

Week 1 — Using Doubles and Near Doubles

Lesson 2

Key Idea

When 10 is added to a number, the digit in the tens place increases by 1. You can use a 99 Chart to add by circling the starting number, using an arrow to show a jump of 10, and shading in the sum.

Try This

Use the Number Construction Mat and Base-Ten Blocks to create a model of the addition problems. Find each sum.

1. $20 + 10 =$ _____
2. $50 + 10 =$ _____
3. $24 + 10 =$ _____
4. $38 + 10 =$ _____

Use a 99 Chart to find each sum.

5. $80 + 10 =$ _____
6. $60 + 10 =$ _____
7. $51 + 10 =$ _____
8. $72 + 10 =$ _____

9. In what direction do you move on a 99 Chart when 10 is added to a number?

_____

Practice

Use the Number Construction Mat and Base-Ten Blocks to create a model of the addition problems. Find each sum.

⑩ 13 + 10 = _____   ⑪ 49 + 10 = _____

⑫ 36 + 10 = _____   ⑬ 65 + 10 = _____

Use a 99 Chart to find each missing addend.

⑭ _____ + 10 = 40   ⑮ _____ + 10 = 61

⑯ _____ + 10 = 66   ⑰ _____ + 10 = 94

Use the Number Construction Mat and Base-Ten Blocks to create a model of the addition problems. Find each missing addend.

⑱ _____ + 10 = 29   ⑲ _____ + 10 = 77

⑳ _____ + 10 = 37   ㉑ _____ + 10 = 55

Reflect

What pattern do you notice when 10 is added to any number?

_____

Using Doubles and Near Doubles • Lesson 2

Week 1 — Using Doubles and Near Doubles

Lesson 3

Key Idea

On a 99 Chart:
Add 10 .... Move UP ↑
Add 1 .... Move RIGHT →
Subtract 1 ...... Move LEFT ←

Try This

Use a 99 Chart to find each sum or difference. Circle the correct direction arrow on the 99 Chart to find the sum or difference.

① 51 + 10 = _____   ↑   →   ←

② 51 + 1 = _____   ↑   →   ←

③ 51 − 1 = _____   ↑   →   ←

④ 47 + 10 = _____   ↑   →   ←

⑤ 47 + 1 = _____   ↑   →   ←

⑥ 47 − 1 = _____   ↑   →   ←

Use a 99 Chart to model each problem. Find each sum. Circle the correct direction arrow or arrows on a 99 Chart to find the sum.

⑦ 58 + 10 = _____   ↑   →   ←

⑧ 63 + 11 =
63 + (10 + 1) = _____   ↑   →   ←

⑨ 15 + 9 =
15 + (10 − 1) = _____   ↑   →   ←

6 Addition • Week 1

Practice

Use a 99 Chart to model each problem. Find each sum and circle the direction or directions that you moved on the 99 Chart.

10   16 + 10 = ____
Start at 16 and move **up or down** 1 block.

11   16 + 9 = ____
Start at 16 and move **up or down** 1 block and **left or right** 1 block.

12   16 + 11 = ____
Start at 16 and move **up or down** 1 block and **left or right** 1 block.

13   34 + 10 = ____
Start at 34 and move **up or down** 1 block.

14   34 + 9 = ____
Start at 34 and move **up or down** 1 block and **left or right** 1 block.

15   34 + 11 = ____
Start at 34 and move **up or down** 1 block and **left or right** 1 block.

Reflect

What do you do differently using a 99 Chart when adding 9 rather than 11 to a number?

_____

_____

_____

Using Doubles and Near Doubles • Lesson 3

Week 1 — Using Doubles and Near Doubles

Lesson 4

Key Idea

10 + 10 = 20

10 + 9 = 19
10 + (10 − 1) = 19
20 − 1 = 19

10 + 11 = 21
10 + (10 + 1) = 21
20 + 1 = 21

Try This
Find the sum of each doubles fact.

1. 3 + 3 = _____
2. 30 + 30 = _____
3. 4 + 4 = _____
4. 40 + 40 = _____

Use a doubles fact to find the sum of each near-doubles fact.

5. 8 + 9 = 8 + (8 + 1)
   _____ + 1 = _____

6. 8 + 7 = 8 + (8 − 1)
   _____ − 1 = _____

7. 20 + 21 = 20 + (20 + 1)
   _____ + 1 = _____

8. 20 + 19 = 20 + (20 − 1)
   _____ − 1 = _____

9. 40 + 41 = 40 + (40 + 1)
   _____ + 1 = _____

10. 40 + 39 = 40 + (40 − 1)
    _____ − 1 = _____

11. 25 + 26 = 25 + (25 + 1)
    _____ + 1 = _____

12. 25 + 24 = 25 + (25 − 1)
    _____ − 1 = _____

Practice
Use the sum of each doubles fact to find the sum of each near-doubles fact.

13 If 30 + 30 = _____, then

30 + 31 = _____.

14 If 15 + 15 = _____, then

15 + 14 = _____.

15 If 30 + 30 = _____, then

30 + 29 = _____.

16 If 15 + 15 = _____, then

15 + 16 = _____.

17 If 50 + 50 = _____, then

50 + 49 = _____.

18 If 45 + 45 = _____, then

45 + 46 = _____.

Use doubles facts to find each missing addend.

19 7 + _____ = 15

20 6 + _____ = 13

21 30 + _____ = 61

22 25 + _____ = 49

Reflect
Explain how using a doubles fact helps you find the sum of a near-doubles fact.

_____

_____

_____

Using Doubles and Near Doubles • Lesson 4

Week 1

Using Doubles and Near Doubles

Lesson 5 Review

This week you explored addition strategies. You discovered that doubles facts are helpful tools when working with near-doubles facts.

Lesson 1 Use a doubles fact to find the sum of each near-doubles fact.

① If 40 + 40 = _____, then 40 + 41 = _____.

② If 25 + 25 = _____, then 25 + 24 = _____.

③ If 35 + 35 = _____, then 35 + 36 = _____.

④ If 50 + 50 = _____, then 49 + 50 = _____.

Lesson 2 Use near-doubles facts to find each missing addend.

⑤ 7 + _____ = 13

⑥ 9 + _____ = 17

⑦ 30 + _____ = 61

⑧ 15 + _____ = 29

Use the Number Construction Mat and Base-Ten Blocks to create a model of the addition problems. Find each sum.

⑨ 16 + 10 = _____

⑩ 10 + 52 = _____

⑪ 43 + 10 = _____

⑫ 88 + 10 = _____

Reflect

Explain how to use a 99 Chart to find the sum of 38 + 9.

_____

_____

Lesson 3   Using a 99 Chart, find each sum.

⑬ 54 + 9 = _____   ⑭ 87 + 9 = _____

⑮ 38 + 11 = _____   ⑯ 19 + 11 = _____

⑰ 22 + 9 = _____   ⑱ 61 + 9 = _____

Use +10 combinations to help you decide if the missing addend is 9 or 11. Circle your choice.

⑲ 15 + (9 or 11) = 26   ⑳ 55 + (9 or 11) = 66

㉑ 34 + (9 or 11) = 43   ㉒ 46 + (9 or 11) = 57

㉓ 79 + (9 or 11) = 88   ㉔ 58 + (9 or 11) = 67

Lesson 4   Using doubles facts, near-doubles facts, +10 combinations, +9 combinations, or +11 combinations, make an addition problem for the following sums. Use a different strategy for each. Write the strategy used on the line provided.

㉕ 54 = _____ + _____   _____

㉖ 48 = _____ + _____   _____

㉗ 66 = _____ + _____   _____

㉘ 39 = _____ + _____   _____

Reflect
Which strategies do you like to use best? Why?

_____

_____

_____

Using Doubles and Near Doubles • Lesson 5 Review

Week 2 — Using Nice Numbers

Lesson 1

Key Idea

6 + 4 = 10

18 + 2 = 20

A **"nice number"** is a number that **ends in a 0**.

Try This

On the number line, circle all the "nice numbers."

1

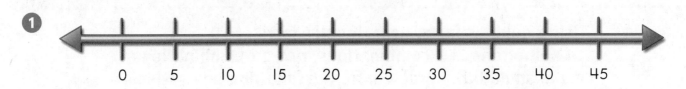

0   5   10   15   20   25   30   35   40   45

Using the number line above, find the given number's closest "nice number."

2 23 _____   **3** 18 _____   **4** 41 _____

5 28 _____   **6** 37 _____   **7** 32 _____

Practice

Make a "nice number" by drawing the missing boxes. Find the missing addend.

8 7 + ____ = 10

9 ____ + 19 = 20

Practice
Make a "nice number" by writing the missing addend.

⑩
_____ + 16 = 20

⑪
_____ + 5 = 10

Use the Number Construction Mat and Base-Ten Blocks to show a model of the following problems. Write an addition sentence using the "nice number" that is described.

⑫ 19 + 3   Share 1 from 3 and add it to 19.

New addition sentence:
_____

⑬ 28 + 6   Share 2 from 6 and add it to 28.

New addition sentence:
_____

Write a new addition sentence by sharing the values. Make sure that a "nice number" is in the new sentence.

⑭ 13 + 6

New addition sentence:
_____

⑮ 34 + 11

New addition sentence:
_____

Reflect
Give three addition problems that have a sum of 24. Then underline the number in each sentence that is closer to a nice number.

_____

_____

_____

Using Nice Numbers • Lesson 1

Using Nice Numbers

Lesson 2

> **Key Idea**
> You can regroup two-digit numbers into "nice numbers" to make them easier to work with.
>
> 24 = 20 + 4
>
> | Tens | Ones |
> |---|---|
> | (2 rods shown) | (4 units shown) |
>
> ↑ Its value is 20.   ↑ Its value is 4.

Try This

Use a Number Construction Mat and Base-Ten Blocks or a Place Value Mat to build each number. Say how many tens and ones are shown.

❶ 24 ____ tens and ____ ones

❷ 12 ____ tens and ____ ones

❸ 35 ____ tens and ____ ones

❹ 29 ____ tens and ____ ones

Use a Number Construction Mat and Base-Ten Blocks or a Place Value Mat to build each addition problem. Say how many tens and ones are shown.

❺ 21 + 11

____ tens and ____ ones = ____

❻ 24 + 15

____ tens and ____ ones = ____

❼ 16 + 13

____ tens and ____ ones = ____

❽ 17 + 12

____ tens and ____ ones = ____

Practice

Regroup each number into tens and ones.

❾ 16 = ____ + ____

❿ 13 = ____ + ____

⓫ 11 = ____ + ____

⓬ 12 = ____ + ____

Regroup each number into tens and ones. Then find each sum.

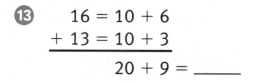

⑬ 16 = 10 + 6
  + 13 = 10 + 3
        20 + 9 = ____

⑭ 22 = 20 + 2
  + 17 = 10 + 7
        ____ + ____ = ____

⑮ 17 = ____ + ____
  + 12 = ____ + ____
        ____ + ____ = ____

⑯ 11 = ____ + ____
  + 24 = ____ + ____
        ____ + ____ = ____

⑰ 13 = ____ + ____
  + 25 = ____ + ____
        ____ + ____ = ____

⑱ 19 = ____ + ____
  + 16 = ____ + ____
        ____ + ____ = ____

⑲ 17 = ____ + ____
  + 11 = ____ + ____
        ____ + ____ = ____

⑳ 27 = ____ + ____
  + 13 = ____ + ____
        ____ + ____ = ____

Reflect
Using a Number Construction Mat and Base-Ten Blocks, draw a model of an addition problem that has a sum of 47.

Tens	Ones

Using Nice Numbers • Lesson 2

Week 2 — Using Nice Numbers

Lesson 3

Key Idea
Regrouping numbers into tens and ones is a strategy for adding two-digit numbers.

$$23 = 20 + 3$$
$$+\ 46 = 40 + 6$$
$$\overline{\qquad\qquad 60 + 9 = 69}$$

Try This
Use a Number Construction Mat and Base-Ten Blocks or a Place Value Mat to build the following numbers. Show how many tens and ones you made.

1 62 _____ tens and _____ ones

2 74 _____ tens and _____ ones

3 48 _____ tens and _____ ones

4 95 _____ tens and _____ ones

Rewrite each number as tens and ones.

5 56 = _____ + _____

6 92 = _____ + _____

7 64 = _____ + _____

8 18 = _____ + _____

Practice
Use the Place Value Mat to build each addition problem. Show how many tens and ones are in the sum.

9 28 + 31

_____ tens and _____ ones = _____

10 52 + 15

_____ tens and _____ ones = _____

11 77 + 12

_____ tens and _____ ones = _____

12 64 + 22

_____ tens and _____ ones = _____

Regroup each number into tens and ones. Then find each sum.

13) 54 = ____ + ____
   + 23 = ____ + ____
   _____
      ____ + ____ = ____

14) 18 = ____ + ____
   + 71 = ____ + ____
   _____
      ____ + ____ = ____

15) 66 = ____ + ____
   + 32 = ____ + ____
   _____
      ____ + ____ = ____

16) 84 = ____ + ____
   + 13 = ____ + ____
   _____
      ____ + ____ = ____

Regroup to find each sum.

17)   62
    + 27
    ____

18)   43
    + 25
    ____

19)   51
    + 18
    ____

20)   23
    + 36
    ____

21)   48
    + 10
    ____

22)   74
    + 26
    ____

Reflect

Jackie thinks it is easier to add 50 + 40 + 6 + 3 than 56 + 43. Do you agree with her? Why or why not?

_____

_____

Using Nice Numbers • Lesson 3

Week 2 — Using Nice Numbers

Lesson 4

Key Idea

"Leaps of 10" is an addition strategy that uses a number line to find the sum.

Use the following steps to help you find sums.
Step 1 Locate the first number on a number line.
Step 2 Break the second number into groups of tens and ones.
Step 3 Make "leaps of 10" and "hops of 1."

Try This

Write how many leaps and hops you make for each number.

① 34   Leaps ____ Hops ____
② 7   Leaps ____ Hops ____
③ 49   Leaps ____ Hops ____
④ 25   Leaps ____ Hops ____
⑤ 36   Leaps ____ Hops ____
⑥ 12   Leaps ____ Hops ____

Practice

Break down the following numbers into tens and ones.

⑦ 52 = ____ tens, ____ ones = 10 + 10 + 10 + 10 + 10 + 1 + 1

⑧ 44 = ____ tens, ____ ones = _____

⑨ 39 = ____ tens, ____ ones = _____

⑩ 21 = ____ tens, ____ ones = _____

Make leaps and hops on the number line to find each sum. Complete the addition sentences.

⑪ 20 + 10 = _____

⑫ 47 + 10 = _____

⑬ 54 + 20 = 54 + _____ + _____ = _____

⑭ 16 + 41 = 16 + _____ + _____ + _____ + _____ + _____ = _____

⑮ 62 + 34 = _____ = _____

⑯ 29 + 22 = _____ = _____

Reflect

Monica says that when you add 42 + 23, you need to make two moves on the number line to find the sum. Is she correct? Explain.

_____

_____

_____

_____

What is the sum of 42 + 23? _____

Using Nice Numbers • Lesson 4

Week 2 — Using Nice Numbers

Lesson 5 Review

This week you explored addition strategies. You used "nice numbers," regrouping, and "leaps of 10" to find the sums.

Lesson 1 Circle the numbers below that are considered "nice numbers."

① 20   31   55   7   94   60   73   68   3   10   19

Write a new addition sentence by sharing the values. Make sure the new sentence includes a "nice number."

② 17 + 8

New addition sentence:

_____

③ 22 + 17

New addition sentence:

_____

Lesson 2 Use a Number Construction Mat and Base-Ten Blocks or a Place Value Mat to build each addition problem. Write how many tens and ones are shown.

④ 41 + 33   ____ tens and ____ ones = ____

⑤ 28 + 31   ____ tens and ____ ones = ____

⑥ 17 + 22   ____ tens and ____ ones = ____

⑦ 63 + 25   ____ tens and ____ ones = ____

Reflect

Explain the sharing you did to write the new addition sentence in Problem 3.

_____

_____

Lesson 3 Regroup each number into tens and ones. Then find each sum.

⑧   43 = 40 + 3
     + 32 = 30 + 2
     _____

⑨   15 = _____
     + 63 = _____
     _____

⑩   13 = _____
     + 25 = _____
     _____

⑪   51 = _____
     + 46 = _____
     _____

Lesson 4 Make leaps and hops on the number line to find each sum.

⑫   23 + 43 = 23 + 10 + 10 + 10 + 10 + 1 + 1 + 1

      = _____

⑬   51 + 34 = 51 + 10 + 10 + 10 + 1 + 1 + 1 + 1 = _____

Reflect

Draw a number line to illustrate the following addition problem using "leaps of 10" and "hops of 1." Explain your moves.

$$56 + 23$$

Week 3 — Using Grouping

Lesson 1

Key Idea
When adding, the order of the numbers does not matter.

$6 + 4 = 10$

$4 + 6 = 10$

Try This
Find each sum.

① $8 + 3 =$ _____ $3 + 8 =$ _____

② $5 + 7 =$ _____ $7 + 5 =$ _____

③ $5 + 4 =$ _____ $4 + 5 =$ _____

④ $20 + 10 =$ _____ $10 + 20 =$ _____

Reorder the numbers in each addition problem to make "nice numbers." Then find the sum.

⑤ $6 + 3 + 4 = 6 + 4 + 3$

_____ + _____ = _____

⑥ $9 + 5 + 1 = 9 + 1 + 5$

_____ + _____ = _____

⑦ $13 + 7 + 4 =$ _____ + _____ + 4

_____ + _____ = _____

Reorder and group the addends to make "nice numbers" to help you find each sum.

⑧ 15 + 3 + 8 + 7 + 5

(15 + ___) + (7 + ___) + 8

___ + ___ + ___ = ___

⑨ 4 + 9 + 16 + 11 + 7

(16 + ___) + (11 + ___) + ___

___ + ___ + ___ = ___

⑩ 21 + 15 + 5 + 7 + 9

(___ + ___) + (___ + ___) + ___

___ + ___ + ___ = ___

⑪ 8 + 13 + 7 + 12 + 5

(___ + ___) + (___ + ___) + ___

___ + ___ + ___ = ___

Reflect
Reorder and group the addends to make the problem easier to solve. Explain why this makes it easier to solve.

18 + 4 + 16 + 10 + 2

_____

_____

_____

_____

Using Grouping • Lesson 1

Week 3 Using Grouping

Lesson 2

Key Idea

Two strategies to make adding easier are the following:

1. Reorder Using Doubles Facts:
   4 + 3 + 4 = (4 + 4) + 3 = 8 + 3 = 11

2. Reorder Using "Nice Numbers":
   2 + 5 + 8 = (8 + 2) + 5 = 10 + 5 = 15

Try This
Write an addition sentence for each Dot Set Card.

1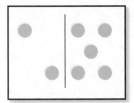

_____

2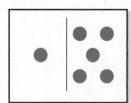

_____

3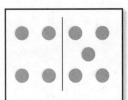

_____

4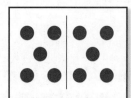

_____

Use the Dot Set Cards to write an addition problem.
Find the sum. Circle the strategy you used.

5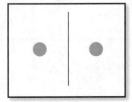

____ + ____ + ____ + ____ = ____

Doubles Fact   or   "Nice Numbers"

6 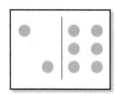 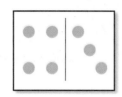   ___ + ___ + ___ + ___ = ___

Doubles Fact   or   "Nice Numbers"

Practice
Use the Dot Set Cards to write an addition problem. Find the sum.

7 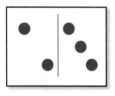 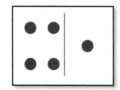   **8**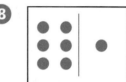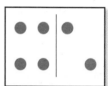

___ + ___ + ___ + ___ = ___   ___ + ___ + ___ + ___ = ___

Use two Dot Set Cards to create an addition sentence that has each sum below. Draw the Dot Set Cards you used to create the sum, and then write the addition sentence below.

9 12 = 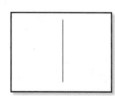   **10** 14 =

___ + ___ + ___ + ___ = 12   ___ + ___ + ___ + ___ = 14

Reflect
Use Dot Set Cards to write two number sentences that have a sum of 15. Create one using the doubles-facts strategy and one using the "nice-numbers" strategy.

1. Doubles Fact: 15 = ___ + ___ + ___ + ___
2. Regrouping with "Nice Numbers":   15 = ___ + ___ + ___ + ___

Using Grouping • Lesson 2

Week 3 — **Using Grouping**

Lesson 3

Key Idea

You can use more than one strategy when adding several numbers.

Doubles Facts	Near-Doubles Facts	"Nice Numbers"
4 + 7 + 4 =	9 + 4 + 5 =	6 + 3 + 4 =
(4 + 4) + 7 =	9 + (5 + 5 − 1) =	(6 + 4) + 3 =
8 + 7 =	9 + (10 − 1) =	10 + 3 =
15	18	13

Try This

Write the sums of both number sentences. Determine whether the calculator input is a correct way of finding it.

1 Original number sentence: 7 + 3 + 5 + 4 = _____

Calculator input: 10 + 9 = _____

Does it work? _____

2 Original number sentence: 4 + 3 + 10 + 4 = _____

Calculator input: 10 + 8 + 4 = _____

Does it work? _____

3 Original number sentence: 8 + 4 + 2 = _____

Calculator input: 10 + 6 = _____

Does it work? _____

4 Original number sentence: 9 + 4 + 7 + 3 = _____

Calculator input: 10 + 3 + 10 = _____

Does it work? _____

⑤ Original number sentence: 4 + 3 + 8 + 3 = ____

Calculator input: 8 + 8 + 2 = ____

Does it work? ____

⑥ Original number sentence: 3 + 8 + 3 + 2 = ____

Calculator input: 10 + 9 = ____

Does it work? ____

Practice
Find each sum. Write the strategy or strategies that you used.

⑦ 10 + 7 + 2 = ____

_____

⑧ 7 + 7 + 2 + 5 = ____

_____

⑨ 3 + 9 + 3 + 9 = ____

_____

⑩ 4 + 6 + 3 + 6 = ____

_____

Reflect
Write a number string of 4 numbers for each sum.

21 _____

18 _____

Week 3 — Using Grouping

Lesson 4

Key Idea
Watch out for numbers in the word problem that are not related to what the question is asking.

Try This
Solve each word problem using addition. Write the number sentence.

① Kevyn went to a carnival three days in a row. On Friday he rode 4 rides. On Saturday he rode 6 rides. On Sunday he rode 4 rides. How many rides did Kevyn ride in all?

_____

② Harold wanted to win tickets from the sports booth. He bought 3 tickets the first round, but he did not win. He bought 5 tickets the second round, but he did not win. Then he bought 7 tickets the third round and finally won. How many total tickets did Harold buy?

_____

③ Elizabeth sold 12 hot dogs during the first hour working at the food stand. During the second hour, she sold 11 hot dogs. During the third and fourth hours, she sold 8 and 10 hot dogs. How many total hot dogs did Elizabeth sell during her four-hour shift?

_____

Practice
Solve each word problem using addition. Write the number sentence.

4. A clown making balloon animals used 4 balloons to make a dog. He used 7 balloons to create a monkey. The giraffe he made required 5 balloons. He made a bird using 4 balloons. How many total balloons did the clown use to make these animals?

_____

5. Margaret started selling tickets for the cake booth at 1:00. She sold the first set in 8 minutes. It took her 5 minutes to sell the second set and 9 minutes to sell the third set. How long did it take for Margaret to sell the three sets of tickets?

_____

Reflect
Angie solved the following problem.

Tyrese bought fish for his fish tank three days in a row this week. On Monday he bought 7 fish. On Tuesday he bought 8 fish. On Wednesday he bought 5 fish. How many fish did Tyrese buy in all?

Angie wrote 3 + 7 + 8 + 5 = 23 fish. Is she correct? If not, what should she have written?

_____

_____

_____

_____

Using Grouping

Lesson 5 Review

This week you explored more addition strategies. You learned that the order of the addends does not affect the sum, so you can reorder the addends to use "nice numbers," doubles facts, and near-doubles facts.

Lesson 1 Answer the following questions.

① Does 8 + 3 = 3 + 8? _____   ② Does 7 + 5 = 5 + 7? _____

If yes, what is the sum? _____   If yes, what is the sum? _____

③ What do you notice about the two problems above?

_____

Lesson 2 Write two number strings for the sum. Use the doubles-facts strategy for one and the "nice-numbers" strategy for the other.

④ 15

Doubles Fact: _____

"Nice Number": _____

Reflect

Show how to reorder the addends to use doubles facts in the problem below. Find the sum. Then show how to reorder the addends to use "nice numbers." Find the sum.

7 + 8 + 3 + 7

_____

_____

Lesson 3 Reorder the addends in the addition problem to make "nice numbers." Then find the sum.

⑤ 6 + 9 + 17 + 14 + 3

(___ + ___) + (___ + ___) + ___

___ + ___ + ___ = ___

Lesson 4 Solve each word problem using addition. Write the number sentence.

⑥ Brittany and Kimberly played **Race for the Sum.** Their first round took 12 minutes. The second round took 16 minutes, and the third round was over in 8 minutes. The final round was finished in 10 minutes. How long did the girls play?

_____

⑦ John was in charge of setting up the carnival booths. There were three rows of booths to set up. The first row took 20 minutes, and the second row took 18 minutes. John needed 22 minutes for the third row. How long did John take to put up all the booths?

_____

Reflect
Explain why addends were reordered in the problem below. Is the sum correct? If not, correct the mistake.

22 + 9 + 13 + 8 + 7   (22 + 8) + (13 + 7) + 9

20 + 20 + 9 = 49

_____

_____

_____

Using Grouping • Lesson 5 Review

Week 4 — Using Partial Sums

Lesson 1

Key Idea
Models and pictures can help with renaming when adding two-digit numbers.

Try This
Circle each set of 10 unit blocks below. Redraw the value given using rods and unit blocks. Write the value.

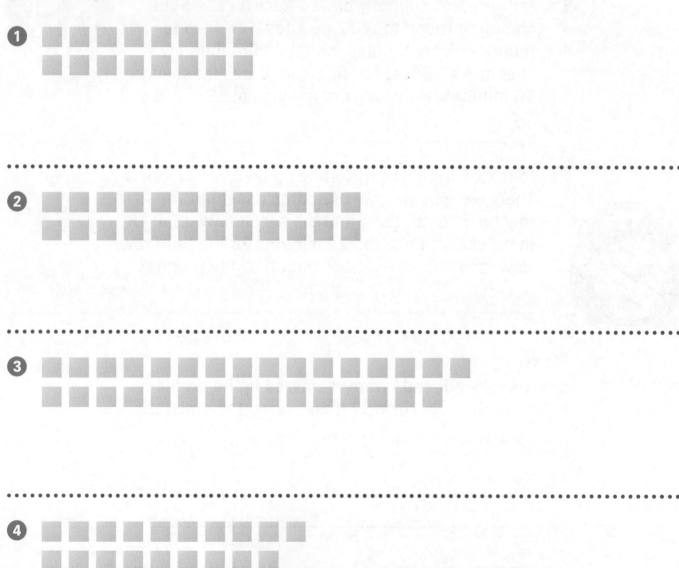

Practice

Use Base-Ten Blocks to create a model for each number. Rename ones to tens as needed. Find the sum.

5) 62
     + 29

6) 45
     + 37

7) 39
     + 14

8) 77
     + 18

Draw a picture using Base-Ten Blocks for each problem. Find the sum.

9) 49
     + 47

10) 67
     + 18

Reflect

Look at the following problem. Is the answer correct? Explain.

```
  59
+ 46
-----
 915
```

_____

_____

_____

Using Partial Sums • Lesson 1

Week 4 — Using Partial Sums

Lesson 2

Key Idea
Models and pictures can help you add three-digit numbers with renaming.

Try This
Circle each set of 10 unit blocks below. Circle each set of 10 rods below. Redraw the value given using flats, rods, and unit blocks. Write the value.

1

2

Practice

Use Base-Ten Blocks to create a model for each number. Rename ones to tens, and tens to hundreds as needed. Find the sum.

3) 137
   + 325

4) 275
   + 142

5) 435
   + 236

6) 378
   + 163

Find the sum.

7) 639
   + 145

8) 434
   + 567

Reflect

Explain the renaming that you did in Problem 8.

_____
_____
_____
_____

Using Partial Sums • Lesson 2

Week 4 — Using Partial Sums

Lesson 3

Key Idea

Find the sum of the tens digits and the sum of the ones digits, and then add them. This strategy is called *partial sums*.

```
   27
  +46
   60   (20 + 40) Add tens.
  +13   (7 + 6) Add ones.
   73
```

Try This

Write each number as the sum of tens and ones values.

1. 46 = _____ + _____
2. 71 = _____ + _____
3. 35 = _____ + _____
4. 62 = _____ + _____

Fill in the missing values using the partial-sums strategy.

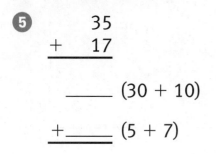

5.
```
      35
   + 17
   ─────
     ___   (30 + 10)
   +___    (5 + 7)
```

6.
```
      64
   + 21
   ─────
     ___   (60 + 20)
   +___    (4 + 1)
```

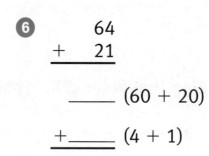

7.
```
      32
   + 56
   ─────
      80  (___ + ___)
   +   8  (___ + ___)
```

8.
```
      49
   + 15
   ─────
      50  (___ + ___)
   + 14  (___ + ___)
```

Practice
Fill in the missing values using the partial-sums strategy.

9)
```
   26
+  43
```
____ (____ + ____)
+____ (____ + ____)

10)
```
   13
+  39
```
____ (____ + ____)
+____ (____ + ____)

Find each sum using the partial-sums strategy.

11)
```
   70
+  22
```

12)
```
   54
+  21
```

13)
```
   28
+  63
```

14)
```
   36
+  43
```

Reflect
Use the partial-sums strategy to find the sum. Explain each of your steps.

```
   54
+  38
```

Using Partial Sums • Lesson 3

Week 4 — Using Partial Sums

Lesson 4

Key Idea
The partial-sums strategy can also be used with three-digit numbers.

```
   314
 + 158
   400   (300 + 100)  Add hundreds.
    60   (10 + 50)    Add tens.
 +  12   (4 + 8)      Add ones.
   472
```

Try This
Write each number as the sum of hundreds, tens, and ones values.

1 182 = _____ + _____ + _____

2 719 = _____ + _____ + _____

3 304 = _____ + _____

4 520 = _____ + _____

Fill in the missing values using the partial-sums strategy.

5
```
    243
  + 151
```
_____ (200 + 100)

_____ (40 + 50)

+_____ (3 + 1)

6
```
    126
  + 315
```
_____ (100 + 300)

_____ (20 + 10)

+_____ (6 + 5)

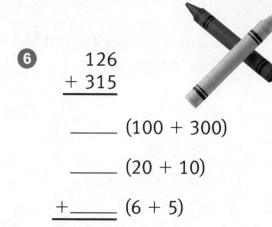

38 Addition • Week 4

Practice
Fill in the missing values using the partial-sums strategy.

7) 436
  + 261

  ____ (____ + ____)
  ____ (____ + ____)
  +____ (____ + ____)

8) 753
  + 138

  ____ (____ + ____)
  ____ (____ + ____)
  +____ (____ + ____)

Find each sum using the partial-sums strategy.

9) 642
  + 153

10) 320
  + 246

11) 343
  + 111

12) 227
  + 516

Reflect
Use the partial-sums strategy to find the sum. Explain each of your steps.

  743
  + 352

_____

_____

_____

Using Partial Sums

Lesson 5 Review

This week you explored more addition strategies. You used partial sums and Base-Ten Blocks with renaming to find sums.

Lesson 1 Use the partial-sums strategy to find each sum.

1   38
    + 55

    _____ (30 + 50)

    +_____ (8 + 5)

2   138
    + 312

    _____ (100 + 300)

    _____ (30 + 10)

    +_____ (8 + 2)

Lesson 2 Use the partial-sums strategy to find each sum.

3   37
    + 44

    _____ (____ + ____)

    +_____ (____ + ____)

4   536
    + 368

    _____ (____ + ____)

    _____ (____ + ____)

    +_____ (____ + ____)

Reflect

Name the partial sums for the problem below. Find the sum.

    556
    + 245

Lesson 3 Use Base-Ten Blocks to create a model for each number. Rename ones to tens and tens to hundreds as needed. Draw your model. Find the sum.

⑤ 64
  + 27

⑥ 34
  + 19

⑦ 78
  + 17

Lesson 4 ## Practice
Use Base-Ten Blocks to create a model for each number. Rename ones to tens and tens to hundreds as needed. Find the sum.

⑧ 177
  + 286

⑨ 454
  + 277

⑩ 194
  + 268

Reflect
Which problem below does not require any renaming to find the sum? Explain.

⑪ 31
  + 55

⑫ 268
  + 513

⑬ 344
  + 292

_____
_____

Using Partial Sums • Lesson 5 Review

Week 1 — Using Doubles and Near Doubles

Practice

Use doubles facts to find the sums of each near-doubles fact.

1. If 20 + 20 = _____, then
   20 + 21 = _____.

2. If 55 + 55 = _____, then
   55 + 56 = _____.

Use doubles facts to find the missing addend.

3. 12 + _____ = 23

4. 14 + _____ = 29

Use the Number Construction Mat and Base-Ten Blocks to create a model of the addition problems. Find each sum.

5. 14 + 10 = _____

6. 46 + 10 = _____

Using a 100 Chart, find each sum.

7. 47 + 9 = _____

8. 19 + 9 = _____

Use +10 combinations to help you decide if the missing addend is 9 or 11. Circle your choice.

9. 18 + (9 or 11) = 29

10. 22 + (9 or 11) = 31

Using doubles facts, near-doubles facts, +10 combinations, +9 combinations, or +11 combinations, make an addition problem for the following sum. Write the strategy used.

11. 39 = _____ + _____   _____

42 Addition • Week 1 Practice

Week 2 — Using Nice Numbers

Practice

Circle the numbers below that are considered "nice numbers."

①   18   40   96   11   29   70   35   10   24   33   64

Write a new addition sentence by sharing the values. Be sure that a "nice number" is in the new sentence.

② 56 + 41

New addition sentence

_____

③ 17 + 62

New addition sentence

_____

Use a Number Construction Mat and Base-Ten Blocks or a Place Value Mat to build each addition problem. Say how many tens and ones are modeled.

④ 23 + 74

____ tens and ____ ones = ____

⑤ 46 + 12

____ tens and ____ ones = ____

⑥ 51 + 18

____ tens and ____ ones = ____

⑦ 36 + 41

____ tens and ____ ones = ____

Regroup each number into tens and ones. Then find each sum.

⑧   42 = 40 + 2
    + 16 = 10 + 6
    _____

⑨   26 = _____
    + 61 = _____
    _____

Addition • Week 2 Practice

Week 3 Using Grouping

Practice

Answer the following questions.

1 Does 6 + 4 = 4 + 6? _____   If yes, what is the sum? _____

2 What do you notice about the problem above?

_____

Reorder the numbers in each addition problem to make "nice numbers." Then find the sum.

3 3 + 6 + 7 = _____   **4** 9 + 8 + 2 + 1 = _____

Reorder the numbers in the addition problem to make "nice numbers." Then find the sum.

5 7 + 5 + 13 + 4 + 5

(____ + ____) + (____ + ____) + ____

____ + ____ + ____ = ____

Solve the word problem using addition. Write the number sentence.

6 Travis sets aside time for reading every night. Four nights ago, Travis read 26 pages of a book. He read 31 pages three nights ago and 21 pages two nights ago. Finally he read 30 pages last night. How many pages did Travis read over the last four nights?

_____

44 Addition • Week 3 Practice

Week 4

Using Partial Sums

Practice

Use the partial-sums strategy to find each sum.

1
```
   284
 + 685
```
_____ (200 + 600)

_____ (80 + 80)

+_____ (4 + 5)

2
```
   158
 + 421
```
500 (_____ + _____)

 70 (_____ + _____)

+ 9 (_____ + _____)

Use the partial-sums strategy to find each sum.

3
```
    46
 +  13
```
_____ (_____ + _____)

+_____ (_____ + _____)

4
```
   824
 + 115
```
_____ (_____ + _____)

_____ (_____ + _____)

+_____ (_____ + _____)

Use Base-Ten Blocks to create a model for each number.
Rename ones to tens and tens to hundreds as needed.
Draw your model. Find the sum.

5
```
    26
 +  37
```

6
```
    14
 +  38
```

7
```
   629
 + 187
```

8
```
   557
 + 385
```

Addition • Week 4 Practice

Unit 3 Workbook

SRAonline.com

Level E

Unit 3 Workbook
Level E

SRA
NUMBER WORLDS
Addition

featuring **Building Blocks** Software

Author
Sharon Griffin
Associate Professor of Education and
Adjunct Associate Professor of Psychology
Clark University
Worcester, Massachusetts

Building Blocks Authors

Douglas H. Clements
Professor of Early Childhood
and Mathematics Education
University at Buffalo
State University of New York, New York

Julie Sarama
Associate Professor of Mathematics Education
University at Buffalo
State University of New York, New York

Contributing Writers
Sherry Booth, *Math Curriculum Developer,* Raleigh, North Carolina
Elizabeth Jimenez, *English Language Learner Consultant,* Pomona, California

Program Reviewers

Jean Delwiche
Almaden Country School
San Jose, California

Cheryl Glorioso
Santa Ana Unified School District
Santa Ana, California

Sharon LaPoint
School District of Indian River County
Vero Beach, Florida

Leigh Lidrbauch
Pasadena Independent School District
Pasadena, Texas

Dave Maresh
Morongo Unified School District
Yucca Valley, California

Mary Mayberry
Mon Valley Education Consortium, AIU 3
Clairton, Pennsylvania

Lauren Parente
Mountain Lakes School District
Mountain Lakes, New Jersey

Juan Regalado
Houston Independent School District
Houston, Texas

M. Kate Thiry
Dublin City School District
Dublin, Ohio

Susan C. Vohrer
Baltimore County Public Schools
Baltimore, Maryland

SRAonline.com

Copyright © 2007 SRA/McGraw-Hill.

All rights reserved. Except as permitted under the United States Copyright Act, no part of this publication may be reproduced or distributed in any form or by any means, or stored in a database or retrieval system, without the prior written permission of the publisher, unless otherwise indicated.

Printed in the United States of America.

Send all inquiries to:
SRA/McGraw-Hill
4400 Easton Commons
Columbus, OH 43219-6188

R5313X.01

9 WCE 12 11 10

Photo Credits
3–39 ©PhotoDisc/Getty Images, Inc.

Contents

Addition

Week 1 Using Doubles and Near Doubles 2

Week 2 Using Nice Numbers 12

Week 3 Using Grouping 22

Week 4 Using Partial Sums 32

Week 1 Practice 42

Week 2 Practice 43

Week 3 Practice 44

Week 4 Practice 45

Week 1 — Using Doubles and Near Doubles

Lesson 1

Key Idea

2 + 2 = 4 Doubles Fact

2 + 3 = 2 + (2 + 1) Near-Doubles Fact

= 4 + 1 = 5

Try This

Find the sum of each doubles fact.

1) 2 + 2 = _____

2) 5 + 5 = _____

3) 4 + 4 = _____

4) 1 + 1 = _____

Use a doubles fact to find the sum of each near-doubles fact.

5) 2 + 3 = 2 + (2 + 1)
_____ + 1 = _____

6) 5 + 6 = 5 + (5 + 1)
_____ + 1 = _____

7) 3 + 4 = 3 + (3 + 1)
_____ + 1 = _____

8) 4 + 5 = 4 + (4 + 1)
_____ + 1 = _____

Practice

Use the sum of each doubles fact to find the sum of each near-doubles fact.

9 If 3 + 3 = _____, then

3 + 4 = _____.

10 If 5 + 5 = _____, then

5 + 6 = _____.

11 If 7 + 7 = _____, then

7 + 8 = _____.

12 If 8 + 8 = _____, then

8 + 9 = _____.

13 If 5 + 5 = _____, then

5 + 4 = _____.

14 If 2 + 2 = _____, then

1 + 2 = _____.

15 If 7 + 7 = _____, then

6 + 7 = _____.

16 If 4 + 4 = _____, then

3 + 4 = _____.

Find each sum.

17 3 + 4 = _____

18 5 + 6 = _____

19 10 + 9 = _____

20 7 + 6 = _____

21 8 + 7 = _____

22 5 + 4 = _____

Reflect

Given the problem 4 + 5, show how each of the doubles facts below can be used to find the sum.

4 + 4   or   5 + 5

_____ or _____

Using Doubles and Near Doubles • Lesson 1

Week 1: Using Doubles and Near Doubles

Lesson 2

Key Idea

When 10 is added to a number, the digit in the tens place increases by 1. You can use a 99 Chart to add by circling the starting number, using an arrow to show a jump of 10, and shading in the sum.

12 + 10 = 22									
21	22	23	24	25	26	27	28	29	30
11	(12)	13	14	15	16	17	18	19	20

Try This

Use the Number Construction Mat and Base-Ten Blocks to create a model of the addition problems. Find each sum.

1. 20 + 10 = _____
2. 50 + 10 = _____
3. 24 + 10 = _____
4. 38 + 10 = _____

Use a 99 Chart to find each sum.

5. 80 + 10 = _____
6. 60 + 10 = _____
7. 51 + 10 = _____
8. 72 + 10 = _____

9. In what direction do you move on a 99 Chart when 10 is added to a number?

_____

Practice

Use the Number Construction Mat and Base-Ten Blocks to create a model of the addition problems. Find each sum.

10 13 + 10 = _____   **11** 49 + 10 = _____

12 36 + 10 = _____   **13** 65 + 10 = _____

Use a 99 Chart to find each missing addend.

14 _____ + 10 = 40   **15** _____ + 10 = 61

16 _____ + 10 = 66   **17** _____ + 10 = 94

Use the Number Construction Mat and Base-Ten Blocks to create a model of the addition problems. Find each missing addend.

18 _____ + 10 = 29   **19** _____ + 10 = 77

20 _____ + 10 = 37   **21** _____ + 10 = 55

Reflect

What pattern do you notice when 10 is added to any number?

_____

Week 1 — Using Doubles and Near Doubles

Lesson 3

Key Idea

On a 99 Chart:
Add 10 .... Move UP ↑
Add 1 .... Move RIGHT →
Subtract 1 ...... Move LEFT ←

Try This

Use a 99 Chart to find each sum or difference. Circle the correct direction arrow on the 99 Chart to find the sum or difference.

1. 51 + 10 = ____   ↑   →   ←

2. 51 + 1 = ____   ↑   →   ←

3. 51 − 1 = ____   ↑   →   ←

4. 47 + 10 = ____   ↑   →   ←

5. 47 + 1 = ____   ↑   →   ←

6. 47 − 1 = ____   ↑   →   ←

Use a 99 Chart to model each problem. Find each sum. Circle the correct direction arrow or arrows on a 99 Chart to find the sum.

7. 58 + 10 = ____   ↑   →   ←

8. 63 + 11 =
   63 + (10 + 1) = ____   ↑   →   ←

9. 15 + 9 =
   15 + (10 − 1) = ____   ↑   →   ←

6 Addition • Week 1

Practice

Use a 99 Chart to model each problem. Find each sum and circle the direction or directions that you moved on the 99 Chart.

⑩ 16 + 10 = _____
Start at 16 and move **up or down** 1 block.

⑪ 16 + 9 = _____
Start at 16 and move **up or down** 1 block and **left or right** 1 block.

⑫ 16 + 11 = _____
Start at 16 and move **up or down** 1 block and **left or right** 1 block.

⑬ 34 + 10 = _____
Start at 34 and move **up or down** 1 block.

⑭ 34 + 9 = _____
Start at 34 and move **up or down** 1 block and **left or right** 1 block.

⑮ 34 + 11 = _____
Start at 34 and move **up or down** 1 block and **left or right** 1 block.

Reflect

What do you do differently using a 99 Chart when adding 9 rather than 11 to a number?

_____
_____
_____

Week 1 — Using Doubles and Near Doubles

Lesson 4

Key Idea

10 + 10 = 20

10 + 9 = 19
10 + (10 − 1) = 19
  \ /
20 − 1 = 19

10 + 11 = 21
10 + (10 + 1) = 21
  \ /
20 + 1 = 21

Try This
Find the sum of each doubles fact.

① 3 + 3 = _____

② 30 + 30 = _____

③ 4 + 4 = _____

④ 40 + 40 = _____

Use a doubles fact to find the sum of each near-doubles fact.

⑤ 8 + 9 = 8 + (8 + 1)
         \ /
    _____ + 1 = _____

⑥ 8 + 7 = 8 + (8 − 1)
         \ /
    _____ − 1 = _____

⑦ 20 + 21 = 20 + (20 + 1)
           \ /
     _____ + 1 = _____

⑧ 20 + 19 = 20 + (20 − 1)
           \ /
     _____ − 1 = _____

⑨ 40 + 41 = 40 + (40 + 1)
           \ /
     _____ + 1 = _____

⑩ 40 + 39 = 40 + (40 − 1)
           \ /
     _____ − 1 = _____

⑪ 25 + 26 = 25 + (25 + 1)
           \ /
     _____ + 1 = _____

⑫ 25 + 24 = 25 + (25 − 1)
           \ /
     _____ − 1 = _____

Practice

Use the sum of each doubles fact to find the sum of each near-doubles fact.

13. If 30 + 30 = _____, then
    30 + 31 = _____.

14. If 15 + 15 = _____, then
    15 + 14 = _____.

15. If 30 + 30 = _____, then
    30 + 29 = _____.

16. If 15 + 15 = _____, then
    15 + 16 = _____.

17. If 50 + 50 = _____, then
    50 + 49 = _____.

18. If 45 + 45 = _____, then
    45 + 46 = _____.

Use doubles facts to find each missing addend.

19. 7 + _____ = 15

20. 6 + _____ = 13

21. 30 + _____ = 61

22. 25 + _____ = 49

Reflect

Explain how using a doubles fact helps you find the sum of a near-doubles fact.

_____

_____

_____

Week 1

Using Doubles and Near Doubles

Lesson 5 Review

This week you explored addition strategies. You discovered that doubles facts are helpful tools when working with near-doubles facts.

Lesson 1 Use a doubles fact to find the sum of each near-doubles fact.

① If 40 + 40 = ____, then
40 + 41 = ____.

② If 25 + 25 = ____, then
25 + 24 = ____.

③ If 35 + 35 = ____, then
35 + 36 = ____.

④ If 50 + 50 = ____, then
49 + 50 = ____.

Lesson 2 Use near-doubles facts to find each missing addend.

⑤ 7 + ____ = 13

⑥ 9 + ____ = 17

⑦ 30 + ____ = 61

⑧ 15 + ____ = 29

Use the Number Construction Mat and Base-Ten Blocks to create a model of the addition problems. Find each sum.

⑨ 16 + 10 = ____

⑩ 10 + 52 = ____

⑪ 43 + 10 = ____

⑫ 88 + 10 = ____

Reflect

Explain how to use a 99 Chart to find the sum of 38 + 9.

_____

_____

Lesson 3   Using a 99 Chart, find each sum.

⑬   54 + 9 = _____   ⑭   87 + 9 = _____

⑮   38 + 11 = _____   ⑯   19 + 11 = _____

⑰   22 + 9 = _____   ⑱   61 + 9 = _____

Use +10 combinations to help you decide if the missing addend is 9 or 11. Circle your choice.

⑲   15 + (9 or 11) = 26   ⑳   55 + (9 or 11) = 66

㉑   34 + (9 or 11) = 43   ㉒   46 + (9 or 11) = 57

㉓   79 + (9 or 11) = 88   ㉔   58 + (9 or 11) = 67

Lesson 4   Using doubles facts, near-doubles facts, +10 combinations, +9 combinations, or +11 combinations, make an addition problem for the following sums. Use a different strategy for each. Write the strategy used on the line provided.

㉕   54 = _____ + _____   _____

㉖   48 = _____ + _____   _____

㉗   66 = _____ + _____   _____

㉘   39 = _____ + _____   _____

Reflect
Which strategies do you like to use best? Why?

_____

_____

_____

Week 2 — Using Nice Numbers

Lesson 1

Key Idea

6 + 4 = 10

18 + 2 = 20

A **"nice number"** is a number that **ends in a 0.**

Try This

On the number line, circle all the "nice numbers."

1

0 5 10 15 20 25 30 35 40 45

Using the number line above, find the given number's closest "nice number."

2 23 _____   **3** 18 _____   **4** 41 _____

5 28 _____   **6** 37 _____   **7** 32 _____

Practice

Make a "nice number" by drawing the missing boxes. Find the missing addend.

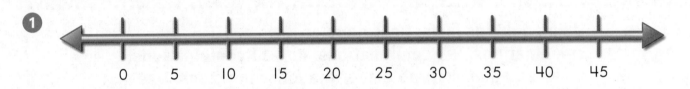

8 7 + ____ = 10

9 ____ + 19 = 20

12 Addition • Week 2

Practice
Make a "nice number" by writing the missing addend.

10 _____ + 16 = 20

11 _____ + 5 = 10

Use the Number Construction Mat and Base-Ten Blocks to show a model of the following problems. Write an addition sentence using the "nice number" that is described.

12 19 + 3   Share 1 from 3 and add it to 19.

New addition sentence:
_____

13 28 + 6   Share 2 from 6 and add it to 28.

New addition sentence:
_____

Write a new addition sentence by sharing the values. Make sure that a "nice number" is in the new sentence.

14 13 + 6

New addition sentence:
_____

15 34 + 11

New addition sentence:
_____

Reflect
Give three addition problems that have a sum of 24. Then underline the number in each sentence that is closer to a nice number.

_____

_____

_____

Using Nice Numbers • Lesson 1

Week 2 — Using Nice Numbers

Lesson 2

Key Idea

You can regroup two-digit numbers into "nice numbers" to make them easier to work with.

24 = 20 + 4

Tens	Ones
(2 tens rods)	(4 ones cubes)

Its value is 20.   Its value is 4.

Try This

Use a Number Construction Mat and Base-Ten Blocks or a Place Value Mat to build each number. Say how many tens and ones are shown.

① 24 _____ tens and _____ ones

② 12 _____ tens and _____ ones

③ 35 _____ tens and _____ ones

④ 29 _____ tens and _____ ones

Use a Number Construction Mat and Base-Ten Blocks or a Place Value Mat to build each addition problem. Say how many tens and ones are shown.

⑤ 21 + 11

_____ tens and _____ ones = _____

⑥ 24 + 15

_____ tens and _____ ones = _____

⑦ 16 + 13

_____ tens and _____ ones = _____

⑧ 17 + 12

_____ tens and _____ ones = _____

Practice

Regroup each number into tens and ones.

⑨ 16 = _____ + _____

⑩ 13 = _____ + _____

⑪ 11 = _____ + _____

⑫ 12 = _____ + _____

Regroup each number into tens and ones.
Then find each sum.

13. 16 = 10 + 6
     + 13 = 10 + 3
     ─────────────
          20 + 9 = ____

14. 22 = 20 + 2
     + 17 = 10 + 7
     ─────────────
          ____ + ____ = ____

15. 17 = ____ + ____
     + 12 = ____ + ____
     ──────────────────
          ____ + ____ = ____

16. 11 = ____ + ____
     + 24 = ____ + ____
     ──────────────────
          ____ + ____ = ____

17. 13 = ____ + ____
     + 25 = ____ + ____
     ──────────────────
          ____ + ____ = ____

18. 19 = ____ + ____
     + 16 = ____ + ____
     ──────────────────
          ____ + ____ = ____

19. 17 = ____ + ____
     + 11 = ____ + ____
     ──────────────────
          ____ + ____ = ____

20. 27 = ____ + ____
     + 13 = ____ + ____
     ──────────────────
          ____ + ____ = ____

Reflect

Using a Number Construction Mat and Base-Ten Blocks, draw a model of an addition problem that has a sum of 47.

Tens	Ones

Week 2 — Lesson 3
Using Nice Numbers

Key Idea
Regrouping numbers into tens and ones is a strategy for adding two-digit numbers.

$$23 = 20 + 3$$
$$+ 46 = 40 + 6$$
$$ 60 + 9 = 69$$

Try This
Use a Number Construction Mat and Base-Ten Blocks or a Place Value Mat to build the following numbers. Show how many tens and ones you made.

① 62 _____ tens and _____ ones

② 74 _____ tens and _____ ones

③ 48 _____ tens and _____ ones

④ 95 _____ tens and _____ ones

Rewrite each number as tens and ones.

⑤ 56 = _____ + _____

⑥ 92 = _____ + _____

⑦ 64 = _____ + _____

⑧ 18 = _____ + _____

Practice
Use the Place Value Mat to build each addition problem. Show how many tens and ones are in the sum.

⑨ 28 + 31

_____ tens and _____ ones = _____

⑩ 52 + 15

_____ tens and _____ ones = _____

⑪ 77 + 12

_____ tens and _____ ones = _____

⑫ 64 + 22

_____ tens and _____ ones = _____

**Regroup each number into tens and ones.
Then find each sum.**

⑬   54 = ____ + ____
   + 23 = ____ + ____
   _____
          ____ + ____ = ____

⑭   18 = ____ + ____
   + 71 = ____ + ____
   _____
          ____ + ____ = ____

⑮   66 = ____ + ____
   + 32 = ____ + ____
   _____
          ____ + ____ = ____

⑯   84 = ____ + ____
   + 13 = ____ + ____
   _____
          ____ + ____ = ____

Regroup to find each sum.

⑰   62
   + 27
   ____

⑱   43
   + 25
   ____

⑲   51
   + 18
   ____

⑳   23
   + 36
   ____

㉑   48
   + 10
   ____

㉒   74
   + 26
   ____

Reflect
Jackie thinks it is easier to add 50 + 40 + 6 + 3 than 56 + 43. Do you agree with her? Why or why not?

_____

_____

Using Nice Numbers • Lesson 3

Week 2 — **Using Nice Numbers**

Lesson 4

Key Idea

"Leaps of 10" is an addition strategy that uses a number line to find the sum.

Use the following steps to help you find sums.
Step 1 Locate the first number on a number line.
Step 2 Break the second number into groups of tens and ones.
Step 3 Make "leaps of 10" and "hops of 1."

Try This
Write how many leaps and hops you make for each number.

① 34   Leaps _____ Hops _____

② 7   Leaps _____ Hops _____

③ 49   Leaps _____ Hops _____

④ 25   Leaps _____ Hops _____

⑤ 36   Leaps _____ Hops _____

⑥ 12   Leaps _____ Hops _____

Practice
Break down the following numbers into tens and ones.

⑦ 52 = _____ tens, _____ ones = 10 + 10 + 10 + 10 + 10 + 1 + 1

⑧ 44 = _____ tens, _____ ones = _____

⑨ 39 = _____ tens, _____ ones = _____

⑩ 21 = _____ tens, _____ ones = _____

Make leaps and hops on the number line to find each sum. Complete the addition sentences.

11 $20 + 10 =$ ____

12 $47 + 10 =$ ____

13 $54 + 20 = 54 +$ ____ $+$ ____ $=$ ____

14 $16 + 41 = 16 +$ ____ $+$ ____ $+$ ____ $+$ ____ $+$ ____ $=$ ____

15 $62 + 34 =$ _____ $=$ ____

16 $29 + 22 =$ _____ $=$ ____

Reflect

Monica says that when you add $42 + 23$, you need to make two moves on the number line to find the sum. Is she correct? Explain.

_____

_____

_____

What is the sum of $42 + 23$? ____

Using Nice Numbers • Lesson 4

Week 2

Using Nice Numbers

Lesson 5 Review

This week you explored addition strategies. You used "nice numbers," regrouping, and "leaps of 10" to find the sums.

Lesson 1 Circle the numbers below that are considered "nice numbers."

❶ 20   31   55   7   94   60   73   68   3   10   19

Write a new addition sentence by sharing the values. Make sure the new sentence includes a "nice number."

❷ 17 + 8

New addition sentence:

_____

❸ 22 + 17

New addition sentence:

_____

Lesson 2 Use a Number Construction Mat and Base-Ten Blocks or a Place Value Mat to build each addition problem. Write how many tens and ones are shown.

❹ 41 + 33   ____ tens and ____ ones = ____

❺ 28 + 31   ____ tens and ____ ones = ____

❻ 17 + 22   ____ tens and ____ ones = ____

❼ 63 + 25   ____ tens and ____ ones = ____

Reflect

Explain the sharing you did to write the new addition sentence in Problem 3.

_____

_____

Lesson 3 Regroup each number into tens and ones. Then find each sum.

⑧ 43 = 40 + 3
  + 32 = 30 + 2
  _____

  _____

⑨   15 = _____
  + 63 = _____

  _____

⑩ 13 = _____
  + 25 = _____

  _____

⑪ 51 = _____
  + 46 = _____

  _____

Lesson 4 Make leaps and hops on the number line to find each sum.

⑫ 23 + 43 = 23 + 10 + 10 + 10 + 10 + 1 + 1 + 1

  = _____

⑬ 51 + 34 = 51 + 10 + 10 + 10 + 1 + 1 + 1 + 1 = _____

Reflect

Draw a number line to illustrate the following addition problem using "leaps of 10" and "hops of 1." Explain your moves.

56 + 23

Week 3 — Using Grouping

Lesson 1

Key Idea

When adding, the order of the numbers does not matter.

1 2 3 4 5 6 7 8 9 10

$6 + 4 = 10$

1 2 3 4 5 6 7 8 9 10

$4 + 6 = 10$

Try This

Find each sum.

① $8 + 3 =$ ____ $3 + 8 =$ ____

② $5 + 7 =$ ____ $7 + 5 =$ ____

③ $5 + 4 =$ ____ $4 + 5 =$ ____

④ $20 + 10 =$ ____ $10 + 20 =$ ____

Reorder the numbers in each addition problem to make "nice numbers." Then find the sum.

⑤ $6 + 3 + 4 = 6 + 4 + 3$

____ + ____ = ____

⑥ $9 + 5 + 1 = 9 + 1 + 5$

____ + ____ = ____

⑦ $13 + 7 + 4 =$ ____ + ____ + 4

____ + ____ = ____

Reorder and group the addends to make "nice numbers" to help you find each sum.

8 $15 + 3 + 8 + 7 + 5$

(15 + ___) + (7 + ___) + 8

___ + ___ + ___ = ___

9 $4 + 9 + 16 + 11 + 7$

(16 + ___) + (11 + ___) + ___

___ + ___ + ___ = ___

10 $21 + 15 + 5 + 7 + 9$

(___ + ___) + (___ + ___) + ___

___ + ___ + ___ = ___

11 $8 + 13 + 7 + 12 + 5$

(___ + ___) + (___ + ___) + ___

___ + ___ + ___ = ___

Reflect

Reorder and group the addends to make the problem easier to solve. Explain why this makes it easier to solve.

$18 + 4 + 16 + 10 + 2$

_____

_____

_____

_____

Using Grouping • Lesson 1

Week 3 Using Grouping
Lesson 2

Key Idea
Two strategies to make adding easier are the following:

1. **Reorder Using Doubles Facts:**
   $4 + 3 + 4 = (4 + 4) + 3 = 8 + 3 = 11$

2. **Reorder Using "Nice Numbers":**
   $2 + 5 + 8 = (8 + 2) + 5 = 10 + 5 = 15$

Try This
Write an addition sentence for each Dot Set Card.

1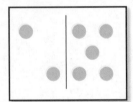

_____

2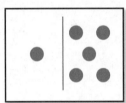

_____

3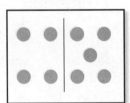

_____

4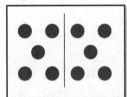

_____

Use the Dot Set Cards to write an addition problem.
Find the sum. Circle the strategy you used.

5

____ + ____ + ____ + ____ = ____

Doubles Fact   or   "Nice Numbers"

24 Addition • Week 3

6 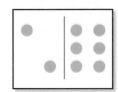 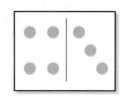 ___ + ___ + ___ + ___ = ___

Doubles Fact   or   "Nice Numbers"

Practice
Use the Dot Set Cards to write an addition problem. Find the sum.

7 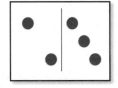    **8**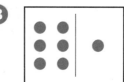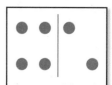

___ + ___ + ___ + ___ = ___   ___ + ___ + ___ + ___ = ___

Use two Dot Set Cards to create an addition sentence that has each sum below. Draw the Dot Set Cards you used to create the sum, and then write the addition sentence below.

9 12 = 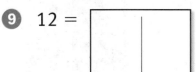   **10** 14 =

___ + ___ + ___ + ___ = 12   ___ + ___ + ___ + ___ = 14

Reflect
Use Dot Set Cards to write two number sentences that have a sum of 15. Create one using the doubles-facts strategy and one using the "nice-numbers" strategy.

1. Doubles Fact: 15 = ___ + ___ + ___ + ___
2. Regrouping with "Nice Numbers":   15 = ___ + ___ + ___ + ___

Week 3 — Using Grouping

Lesson 3

Key Idea

You can use more than one strategy when adding several numbers.

Doubles Facts	Near-Doubles Facts	"Nice Numbers"
4 + 7 + 4 =	9 + 4 + 5 =	6 + 3 + 4 =
(4 + 4) + 7 =	9 + (5 + 5 − 1) =	(6 + 4) + 3 =
8 + 7 =	9 + (10 − 1) =	10 + 3 =
15	18	13

Try This

Write the sums of both number sentences. Determine whether the calculator input is a correct way of finding it.

❶ Original number sentence: 7 + 3 + 5 + 4 = _____

Calculator input: 10 + 9 = _____

Does it work? _____

❷ Original number sentence: 4 + 3 + 10 + 4 = _____

Calculator input: 10 + 8 + 4 = _____

Does it work? _____

❸ Original number sentence: 8 + 4 + 2 = _____

Calculator input: 10 + 6 = _____

Does it work? _____

❹ Original number sentence: 9 + 4 + 7 + 3 = _____

Calculator input: 10 + 3 + 10 = _____

Does it work? _____

5 Original number sentence: 4 + 3 + 8 + 3 = _____

   Calculator input: 8 + 8 + 2 = _____

   Does it work?   _____

6 Original number sentence: 3 + 8 + 3 + 2 = _____

   Calculator input: 10 + 9 = _____

   Does it work?   _____

Practice
Find each sum. Write the strategy or strategies that you used.

7 10 + 7 + 2 = _____

_____

8 7 + 7 + 2 + 5 = _____

_____

9 3 + 9 + 3 + 9 = _____

_____

10 4 + 6 + 3 + 6 = _____

_____

Reflect
Write a number string of 4 numbers for each sum.

21 _____

18 _____

Week 3 — Using Grouping

Lesson 4

Key Idea
Watch out for numbers in the word problem that are not related to what the question is asking.

Try This
Solve each word problem using addition. Write the number sentence.

1. Kevyn went to a carnival three days in a row. On Friday he rode 4 rides. On Saturday he rode 6 rides. On Sunday he rode 4 rides. How many rides did Kevyn ride in all?

   _____

2. Harold wanted to win tickets from the sports booth. He bought 3 tickets the first round, but he did not win. He bought 5 tickets the second round, but he did not win. Then he bought 7 tickets the third round and finally won. How many total tickets did Harold buy?

   _____

3. Elizabeth sold 12 hot dogs during the first hour working at the food stand. During the second hour, she sold 11 hot dogs. During the third and fourth hours, she sold 8 and 10 hot dogs. How many total hot dogs did Elizabeth sell during her four-hour shift?

   _____

Practice
Solve each word problem using addition. Write the number sentence.

④ A clown making balloon animals used 4 balloons to make a dog. He used 7 balloons to create a monkey. The giraffe he made required 5 balloons. He made a bird using 4 balloons. How many total balloons did the clown use to make these animals?

_____

⑤ Margaret started selling tickets for the cake booth at 1:00. She sold the first set in 8 minutes. It took her 5 minutes to sell the second set and 9 minutes to sell the third set. How long did it take for Margaret to sell the three sets of tickets?

_____

Reflect
Angie solved the following problem.

Tyrese bought fish for his fish tank three days in a row this week. On Monday he bought 7 fish. On Tuesday he bought 8 fish. On Wednesday he bought 5 fish. How many fish did Tyrese buy in all?

Angie wrote $3 + 7 + 8 + 5 = 23$ fish. Is she correct? If not, what should she have written?

_____

_____

_____

_____

Using Grouping

Lesson 5 Review

This week you explored more addition strategies. You learned that the order of the addends does not affect the sum, so you can reorder the addends to use "nice numbers," doubles facts, and near-doubles facts.

Lesson 1   Answer the following questions.

① Does 8 + 3 = 3 + 8? _____   ② Does 7 + 5 = 5 + 7? _____

If yes, what is the sum? _____   If yes, what is the sum? _____

③ What do you notice about the two problems above?

_____

Lesson 2   Write two number strings for the sum. Use the doubles-facts strategy for one and the "nice-numbers" strategy for the other.

④ 15

Doubles Fact: _____

"Nice Number": _____

Reflect

Show how to reorder the addends to use doubles facts in the problem below. Find the sum. Then show how to reorder the addends to use "nice numbers." Find the sum.

7 + 8 + 3 + 7

_____

_____

Lesson 3 — Reorder the addends in the addition problem to make "nice numbers." Then find the sum.

⑤ 6 + 9 + 17 + 14 + 3

(___ + ___) + (___ + ___) + ___

___ + ___ + ___ = ___

Lesson 4 — Solve each word problem using addition. Write the number sentence.

⑥ Brittany and Kimberly played **Race for the Sum**. Their first round took 12 minutes. The second round took 16 minutes, and the third round was over in 8 minutes. The final round was finished in 10 minutes. How long did the girls play?

⑦ John was in charge of setting up the carnival booths. There were three rows of booths to set up. The first row took 20 minutes, and the second row took 18 minutes. John needed 22 minutes for the third row. How long did John take to put up all the booths?

Reflect

Explain why addends were reordered in the problem below. Is the sum correct? If not, correct the mistake.

22 + 9 + 13 + 8 + 7   (22 + 8) + (13 + 7) + 9

20 + 20 + 9 = 49

___

___

___

Using Grouping • Lesson 5 Review

Week 4 — Using Partial Sums

Lesson 1

Key Idea
Models and pictures can help with renaming when adding two-digit numbers.

Try This
Circle each set of 10 unit blocks below. Redraw the value given using rods and unit blocks. Write the value.

1.

2.

3.

4.

32 Addition • Week 4

Practice

Use Base-Ten Blocks to create a model for each number. Rename ones to tens as needed. Find the sum.

⑤   62
   + 29

⑥   45
   + 37

⑦   39
   + 14

⑧   77
   + 18

Draw a picture using Base-Ten Blocks for each problem. Find the sum.

⑨   49
   + 47

· ·

⑩   67
   + 18

Reflect

Look at the following problem. Is the answer correct? Explain.

   59
 + 46
 ────
  915

_____

_____

_____

Using Partial Sums • Lesson 1   33

Week 4 — Using Partial Sums

Lesson 2

> **Key Idea**
> Models and pictures can help you add three-digit numbers with renaming.

Try This

Circle each set of 10 unit blocks below. Circle each set of 10 rods below. Redraw the value given using flats, rods, and unit blocks. Write the value.

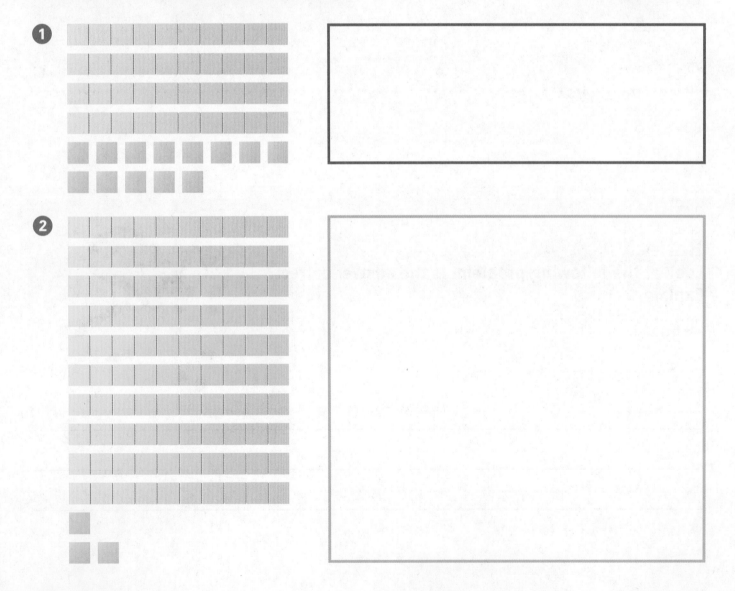

Practice
Use Base-Ten Blocks to create a model for each number. Rename ones to tens, and tens to hundreds as needed. Find the sum.

3) 137
  + 325

4) 275
  + 142

5) 435
  + 236

6) 378
  + 163

Find the sum.

7) 639
  + 145

8) 434
  + 567

Reflect
Explain the renaming that you did in Problem 8.

_____
_____
_____
_____

Week 4 — Using Partial Sums

Lesson 3

Key Idea

Find the sum of the tens digits and the sum of the ones digits, and then add them. This strategy is called *partial sums*.

```
   27
  +46
   60   (20 + 40) Add tens.
  +13   (7 + 6) Add ones.
   73
```

Try This

Write each number as the sum of tens and ones values.

1 46 = _____ + _____

2 71 = _____ + _____

3 35 = _____ + _____

4 62 = _____ + _____

Fill in the missing values using the partial-sums strategy.

5
```
    35
+   17
─────
_____  (30 + 10)
+_____ (5 + 7)
```

6
```
    64
+   21
─────
_____  (60 + 20)
+_____ (4 + 1)
```

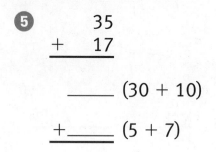

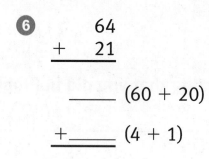

7
```
    32
+   56
─────
    80  (____ + ____)
+    8  (____ + ____)
```

8
```
    49
+   15
─────
    50  (____ + ____)
+   14  (____ + ____)
```

36 Addition • Week 4

Practice

Fill in the missing values using the partial-sums strategy.

9)
```
    26
+   43
```
____ (____ + ____)
+____ (____ + ____)

10)
```
    13
+   39
```
____ (____ + ____)
+____ (____ + ____)

Find each sum using the partial-sums strategy.

11)
```
    70
+   22
```

12)
```
    54
+   21
```

13)
```
    28
+   63
```

14)
```
    36
+   43
```

Reflect

Use the partial-sums strategy to find the sum. Explain each of your steps.

```
    54
+   38
```

_____

_____

_____

Using Partial Sums • Lesson 3

Week 4 — Using Partial Sums

Lesson 4

Key Idea
The partial-sums strategy can also be used with three-digit numbers.

```
    314
  + 158
  -----
    400   (300 + 100) Add hundreds.
     60   (10 + 50) Add tens.
  +  12   (4 + 8) Add ones.
  -----
    472
```

Try This
Write each number as the sum of hundreds, tens, and ones values.

1) 182 = ____ + ____ + ____

2) 719 = ____ + ____ + ____

3) 304 = ____ + ____

4) 520 = ____ + ____

Fill in the missing values using the partial-sums strategy.

5)
```
    243
  + 151
  -----
  ____  (200 + 100)
  ____  (40 + 50)
 +____  (3 + 1)
```

6)
```
    126
  + 315
  -----
  ____  (100 + 300)
  ____  (20 + 10)
 +____  (6 + 5)
```

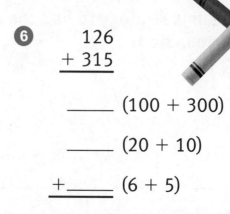

38 Addition • Week 4

Practice
Fill in the missing values using the partial-sums strategy.

7.   436
   + 261
   ___ (___ + ___)
   ___ (___ + ___)
   + ___ (___ + ___)

8.   753
   + 138
   ___ (___ + ___)
   ___ (___ + ___)
   + ___ (___ + ___)

Find each sum using the partial-sums strategy.

9.   642
   + 153

10.   320
   + 246

11.   343
   + 111

12.   227
   + 516

Reflect
Use the partial-sums strategy to find the sum. Explain each of your steps.

   743
   + 352

_____

_____

_____

Using Partial Sums • Lesson 4

Week 4 — Using Partial Sums

Lesson 5 Review

This week you explored more addition strategies. You used partial sums and Base-Ten Blocks with renaming to find sums.

Lesson 1 Use the partial-sums strategy to find each sum.

1
```
   38
+  55
```
____ (30 + 50)

+____ (8 + 5)

2
```
   138
+  312
```
____ (100 + 300)

____ (30 + 10)

+____ (8 + 2)

Lesson 2 Use the partial-sums strategy to find each sum.

3
```
   37
+  44
```
____ (____ + ____)

+____ (____ + ____)

4
```
   536
+  368
```
____ (____ + ____)

____ (____ + ____)

+____ (____ + ____)

Reflect
Name the partial sums for the problem below.
Find the sum.

```
   556
+  245
```

40 Addition • Week 4

Lesson 3 Use Base-Ten Blocks to create a model for each number. Rename ones to tens and tens to hundreds as needed. Draw your model. Find the sum.

⑤   64
     + 27

⑥   34
     + 19

⑦   78
     + 17

Lesson 4

Practice
Use Base-Ten Blocks to create a model for each number. Rename ones to tens and tens to hundreds as needed. Find the sum.

⑧   177
     + 286

⑨   454
     + 277

⑩   194
     + 268

Reflect
Which problem below does not require any renaming to find the sum? Explain.

⑪   31
     + 55

⑫   268
     + 513

⑬   344
     + 292

_____

_____

Using Partial Sums • Lesson 5 Review

Week 1 — Using Doubles and Near Doubles

Practice

Use doubles facts to find the sums of each near-doubles fact.

① If 20 + 20 = _____, then
20 + 21 = _____.

② If 55 + 55 = _____, then
55 + 56 = _____.

Use doubles facts to find the missing addend.

③ 12 + _____ = 23

④ 14 + _____ = 29

Use the Number Construction Mat and Base-Ten Blocks to create a model of the addition problems. Find each sum.

⑤ 14 + 10 = _____

⑥ 46 + 10 = _____

Using a 100 Chart, find each sum.

⑦ 47 + 9 = _____

⑧ 19 + 9 = _____

Use +10 combinations to help you decide if the missing addend is 9 or 11. Circle your choice.

⑨ 18 + (9 or 11) = 29

⑩ 22 + (9 or 11) = 31

Using doubles facts, near-doubles facts, +10 combinations, +9 combinations, or +11 combinations, make an addition problem for the following sum. Write the strategy used.

⑪ 39 = _____ + _____   _____

42 Addition • Week 1 Practice

Week 2 Using Nice Numbers

Practice

Circle the numbers below that are considered "nice numbers."

1 18   40   96   11   29   70   35   10   24   33   64

Write a new addition sentence by sharing the values. Be sure that a "nice number" is in the new sentence.

2 56 + 41

New addition sentence

_____

3 17 + 62

New addition sentence

_____

Use a Number Construction Mat and Base-Ten Blocks or a Place Value Mat to build each addition problem. Say how many tens and ones are modeled.

4 23 + 74

_____ tens and _____ ones = _____

5 46 + 12

_____ tens and _____ ones = _____

6 51 + 18

_____ tens and _____ ones = _____

7 36 + 41

_____ tens and _____ ones = _____

Regroup each number into tens and ones. Then find each sum.

8   42 = 40 + 2
     + 16 = 10 + 6
     _____

9   26 = _____
     + 61 = _____
     _____

Addition • Week 2 Practice   **43**

Week 3 Using Grouping

Practice

Answer the following questions.

① Does 6 + 4 = 4 + 6? _____   If yes, what is the sum? _____

② What do you notice about the problem above?

_____

Reorder the numbers in each addition problem to make "nice numbers." Then find the sum.

③ 3 + 6 + 7 = _____   ④ 9 + 8 + 2 + 1 = _____

Reorder the numbers in the addition problem to make "nice numbers." Then find the sum.

⑤ 7 + 5 + 13 + 4 + 5

(____ + ____) + (____ + ____) + ____

____ + ____ + ____ = ____

Solve the word problem using addition. Write the number sentence.

⑥ Travis sets aside time for reading every night. Four nights ago, Travis read 26 pages of a book. He read 31 pages three nights ago and 21 pages two nights ago. Finally he read 30 pages last night. How many pages did Travis read over the last four nights?

_____

44 Addition • Week 3 Practice

Week 4: Using Partial Sums

Practice

Use the partial-sums strategy to find each sum.

1) 284
　　+ 685

　　____ (200 + 600)

　　____ (80 + 80)

　+____ (4 + 5)
　____

2) 158
　　+ 421

　　500 (____ + ____)

　　 70 (____ + ____)

　+ 9 (____ + ____)

Use the partial-sums strategy to find each sum.

3) 46
　　+ 13

　　____ (____ + ____)

　+____ (____ + ____)

4) 824
　　+ 115

　　____ (____ + ____)

　　____ (____ + ____)

　+____ (____ + ____)

Use Base-Ten Blocks to create a model for each number. Rename ones to tens and tens to hundreds as needed. Draw your model. Find the sum.

5) 26
　　+ 37

6) 14
　　+ 38

7) 629
　　+ 187

8) 557
　　+ 385

Addition • Week 4 Practice **45**

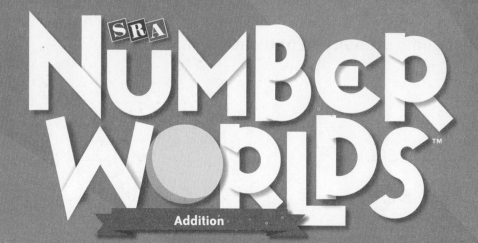

Unit 3 Workbook

SRAonline.com

Level E

SRA Number Worlds

Addition

Unit 3 Workbook
Level E

featuring Building Blocks Software

Author
Sharon Griffin
*Associate Professor of Education and
Adjunct Associate Professor of Psychology*
Clark University
Worcester, Massachusetts

Building Blocks Authors

Douglas H. Clements
*Professor of Early Childhood
and Mathematics Education*
University at Buffalo
State University of New York, New York

Julie Sarama
Associate Professor of Mathematics Education
University at Buffalo
State University of New York, New York

Contributing Writers
Sherry Booth, *Math Curriculum Developer,* Raleigh, North Carolina
Elizabeth Jimenez, *English Language Learner Consultant,* Pomona, California

Program Reviewers

Jean Delwiche
Almaden Country School
San Jose, California

Cheryl Glorioso
Santa Ana Unified School District
Santa Ana, California

Sharon LaPoint
School District of Indian River County
Vero Beach, Florida

Leigh Lidrbauch
Pasadena Independent School District
Pasadena, Texas

Dave Maresh
Morongo Unified School District
Yucca Valley, California

Mary Mayberry
Mon Valley Education Consortium, AIU 3
Clairton, Pennsylvania

Lauren Parente
Mountain Lakes School District
Mountain Lakes, New Jersey

Juan Regalado
Houston Independent School District
Houston, Texas

M. Kate Thiry
Dublin City School District
Dublin, Ohio

Susan C. Vohrer
Baltimore County Public Schools
Baltimore, Maryland

SRAonline.com

Copyright © 2007 SRA/McGraw-Hill.

All rights reserved. Except as permitted under the United States Copyright Act, no part of this publication may be reproduced or distributed in any form or by any means, or stored in a database or retrieval system, without the prior written permission of the publisher, unless otherwise indicated.

Printed in the United States of America.

Send all inquiries to:
SRA/McGraw-Hill
4400 Easton Commons
Columbus, OH 43219-6188

R5313X.01

9 WCE 12 11 10

Photo Credits
3-39 ©PhotoDisc/Getty Images, Inc.

Contents

Addition

Week 1 Using Doubles and Near Doubles 2

Week 2 Using Nice Numbers 12

Week 3 Using Grouping 22

Week 4 Using Partial Sums 32

Week 1 Practice 42

Week 2 Practice 43

Week 3 Practice 44

Week 4 Practice 45

Week 1 — Using Doubles and Near Doubles

Lesson 1

Key Idea

2	+	2	=	4 Doubles Fact
2	+	3	=	2 + (2 + 1) Near-Doubles Fact
			=	4 + 1 = 5

Try This
Find the sum of each doubles fact.

① 2 + 2 = _____

② 5 + 5 = _____

③ 4 + 4 = _____

④ 1 + 1 = _____

Use a doubles fact to find the sum of each near-doubles fact.

⑤ 2 + 3 = 2 + (2 + 1)
   _____ + 1 = _____

⑥ 5 + 6 = 5 + (5 + 1)
   _____ + 1 = _____

⑦ 3 + 4 = 3 + (3 + 1)
   _____ + 1 = _____

⑧ 4 + 5 = 4 + (4 + 1)
   _____ + 1 = _____

Practice

Use the sum of each doubles fact to find the sum of each near-doubles fact.

9 If 3 + 3 = ____, then

3 + 4 = ____.

10 If 5 + 5 = ____, then

5 + 6 = ____.

11 If 7 + 7 = ____, then

7 + 8 = ____.

12 If 8 + 8 = ____, then

8 + 9 = ____.

13 If 5 + 5 = ____, then

5 + 4 = ____.

14 If 2 + 2 = ____, then

1 + 2 = ____.

15 If 7 + 7 = ____, then

6 + 7 = ____.

16 If 4 + 4 = ____, then

3 + 4 = ____.

Find each sum.

17 3 + 4 = ____

18 5 + 6 = ____

19 10 + 9 = ____

20 7 + 6 = ____

21 8 + 7 = ____

22 5 + 4 = ____

Reflect

Given the problem 4 + 5, show how each of the doubles facts below can be used to find the sum.

    4 + 4   or   5 + 5

_____ or _____

Using Doubles and Near Doubles • Lesson 1

Week 1 — Using Doubles and Near Doubles

Lesson 2

Key Idea

When 10 is added to a number, the digit in the tens place increases by 1. You can use a 99 Chart to add by circling the starting number, using an arrow to show a jump of 10, and shading in the sum.

12 + 10 = 22

21	22	23	24	25	26	27	28	29	30
11	12	13	14	15	16	17	18	19	20

Try This

Use the Number Construction Mat and Base-Ten Blocks to create a model of the addition problems. Find each sum.

1. 20 + 10 = _____
2. 50 + 10 = _____
3. 24 + 10 = _____
4. 38 + 10 = _____

Use a 99 Chart to find each sum.

5. 80 + 10 = _____
6. 60 + 10 = _____
7. 51 + 10 = _____
8. 72 + 10 = _____

9. In what direction do you move on a 99 Chart when 10 is added to a number?

_____

Practice
Use the Number Construction Mat and Base-Ten Blocks to create a model of the addition problems. Find each sum.

10 13 + 10 = ____   **11** 49 + 10 = ____

12 36 + 10 = ____   **13** 65 + 10 = ____

Use a 99 Chart to find each missing addend.

14 ____ + 10 = 40   **15** ____ + 10 = 61

16 ____ + 10 = 66   **17** ____ + 10 = 94

Use the Number Construction Mat and Base-Ten Blocks to create a model of the addition problems. Find each missing addend.

18 ____ + 10 = 29   **19** ____ + 10 = 77

20 ____ + 10 = 37   **21** ____ + 10 = 55

Reflect
What pattern do you notice when 10 is added to any number?

_____

Week 1 — Using Doubles and Near Doubles

Lesson 3

Key Idea

On a 99 Chart:
Add 10 .... Move UP ↑
Add 1 .... Move RIGHT →
Subtract 1 ...... Move LEFT ←

Try This

Use a 99 Chart to find each sum or difference. Circle the correct direction arrow on the 99 Chart to find the sum or difference.

1. $51 + 10 =$ _____   ↑ → ←

2. $51 + 1 =$ _____   ↑ → ←

3. $51 - 1 =$ _____   ↑ → ←

4. $47 + 10 =$ _____   ↑ → ←

5. $47 + 1 =$ _____   ↑ → ←

6. $47 - 1 =$ _____   ↑ → ←

Use a 99 Chart to model each problem. Find each sum. Circle the correct direction arrow or arrows on a 99 Chart to find the sum.

7. $58 + 10 =$ _____   ↑ → ←

8. $63 + 11 =$
   $63 + (10 + 1) =$ _____   ↑ → ←

9. $15 + 9 =$
   $15 + (10 - 1) =$ _____   ↑ → ←

6 Addition • Week 1

Practice

Use a 99 Chart to model each problem. Find each sum and circle the direction or directions that you moved on the 99 Chart.

⑩ 16 + 10 = _____
Start at 16 and move **up or down** 1 block.

⑪ 16 + 9 = _____
Start at 16 and move **up or down** 1 block and **left or right** 1 block.

⑫ 16 + 11 = _____
Start at 16 and move **up or down** 1 block and **left or right** 1 block.

⑬ 34 + 10 = _____
Start at 34 and move **up or down** 1 block.

⑭ 34 + 9 = _____
Start at 34 and move **up or down** 1 block and **left or right** 1 block.

⑮ 34 + 11 = _____
Start at 34 and move **up or down** 1 block and **left or right** 1 block.

Reflect

What do you do differently using a 99 Chart when adding 9 rather than 11 to a number?

_____

_____

_____

Using Doubles and Near Doubles • Lesson 3

Week 1 — Using Doubles and Near Doubles

Lesson 4

Key Idea

10 + 10 = 20

10 + 9 = 19
10 + (10 − 1) = 19
20 − 1 = 19

10 + 11 = 21
10 + (10 + 1) = 21
20 + 1 = 21

Try This
Find the sum of each doubles fact.

① 3 + 3 = _____

② 30 + 30 = _____

③ 4 + 4 = _____

④ 40 + 40 = _____

Use a doubles fact to find the sum of each near-doubles fact.

⑤ 8 + 9 = 8 + (8 + 1)
_____ + 1 = _____

⑥ 8 + 7 = 8 + (8 − 1)
_____ − 1 = _____

⑦ 20 + 21 = 20 + (20 + 1)
_____ + 1 = _____

⑧ 20 + 19 = 20 + (20 − 1)
_____ − 1 = _____

⑨ 40 + 41 = 40 + (40 + 1)
_____ + 1 = _____

⑩ 40 + 39 = 40 + (40 − 1)
_____ − 1 = _____

⑪ 25 + 26 = 25 + (25 + 1)
_____ + 1 = _____

⑫ 25 + 24 = 25 + (25 − 1)
_____ − 1 = _____

8 Addition • Week 1

Practice
Use the sum of each doubles fact to find the sum of each near-doubles fact.

⑬ If 30 + 30 = ____, then
30 + 31 = ____.

⑭ If 15 + 15 = ____, then
15 + 14 = ____.

⑮ If 30 + 30 = ____, then
30 + 29 = ____.

⑯ If 15 + 15 = ____, then
15 + 16 = ____.

⑰ If 50 + 50 = ____, then
50 + 49 = ____.

⑱ If 45 + 45 = ____, then
45 + 46 = ____.

Use doubles facts to find each missing addend.

⑲ 7 + ____ = 15

⑳ 6 + ____ = 13

㉑ 30 + ____ = 61

㉒ 25 + ____ = 49

Reflect
Explain how using a doubles fact helps you find the sum of a near-doubles fact.

_____

_____

_____

Week 1

Using Doubles and Near Doubles

Lesson 5 Review

This week you explored addition strategies. You discovered that doubles facts are helpful tools when working with near-doubles facts.

Lesson 1 Use a doubles fact to find the sum of each near-doubles fact.

1. If 40 + 40 = _____, then 40 + 41 = _____.

2. If 25 + 25 = _____, then 25 + 24 = _____.

3. If 35 + 35 = _____, then 35 + 36 = _____.

4. If 50 + 50 = _____, then 49 + 50 = _____.

Lesson 2 Use near-doubles facts to find each missing addend.

5. 7 + _____ = 13

6. 9 + _____ = 17

7. 30 + _____ = 61

8. 15 + _____ = 29

Use the Number Construction Mat and Base-Ten Blocks to create a model of the addition problems. Find each sum.

9. 16 + 10 = _____

10. 10 + 52 = _____

11. 43 + 10 = _____

12. 88 + 10 = _____

Reflect

Explain how to use a 99 Chart to find the sum of 38 + 9.

_____

_____

Lesson 3   Using a 99 Chart, find each sum.

⑬ 54 + 9 = _____   ⑭ 87 + 9 = _____

⑮ 38 + 11 = _____   ⑯ 19 + 11 = _____

⑰ 22 + 9 = _____   ⑱ 61 + 9 = _____

Use +10 combinations to help you decide if the missing addend is 9 or 11. Circle your choice.

⑲ 15 + (9 or 11) = 26   ⑳ 55 + (9 or 11) = 66

㉑ 34 + (9 or 11) = 43   ㉒ 46 + (9 or 11) = 57

㉓ 79 + (9 or 11) = 88   ㉔ 58 + (9 or 11) = 67

Lesson 4   Using doubles facts, near-doubles facts, +10 combinations, +9 combinations, or +11 combinations, make an addition problem for the following sums. Use a different strategy for each. Write the strategy used on the line provided.

㉕ 54 = _____ + _____   _____

㉖ 48 = _____ + _____   _____

㉗ 66 = _____ + _____   _____

㉘ 39 = _____ + _____   _____

Reflect
Which strategies do you like to use best? Why?

_____

_____

_____

Week 2 — Lesson 1: Using Nice Numbers

Key Idea

$6 + 4 = 10$

$18 + 2 = 20$

A **"nice number"** is a number that **ends in a 0**.

Try This
On the number line, circle all the "nice numbers."

1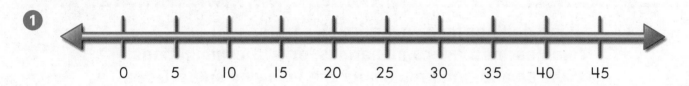

Using the number line above, find the given number's closest "nice number."

2 23 _____   **3** 18 _____   **4** 41 _____

5 28 _____   **6** 37 _____   **7** 32 _____

Practice
Make a "nice number" by drawing the missing boxes.
Find the missing addend.

8 $7 + \_\_\_ = 10$

9 $\_\_\_ + 19 = 20$

12 Addition • Week 2

Practice
Make a "nice number" by writing the missing addend.

⑩

____ + 16 = 20

⑪ ____ + 5 = 10

Use the Number Construction Mat and Base-Ten Blocks to show a model of the following problems. Write an addition sentence using the "nice number" that is described.

⑫ 19 + 3   Share 1 from 3 and add it to 19.

New addition sentence:
_____

⑬ 28 + 6   Share 2 from 6 and add it to 28.

New addition sentence:
_____

Write a new addition sentence by sharing the values. Make sure that a "nice number" is in the new sentence.

⑭ 13 + 6

New addition sentence:
_____

⑮ 34 + 11

New addition sentence:
_____

Reflect
Give three addition problems that have a sum of 24. Then underline the number in each sentence that is closer to a nice number.

_____

_____

_____

Week 2

Using Nice Numbers

Lesson 2

Key Idea

You can regroup two-digit numbers into "nice numbers" to make them easier to work with.

24 = 20 + 4

Tens	Ones

↑ Its value is 20.   ↑ Its value is 4.

Try This

Use a Number Construction Mat and Base-Ten Blocks or a Place Value Mat to build each number. Say how many tens and ones are shown.

1 24 _____ tens and _____ ones

2 12 _____ tens and _____ ones

3 35 _____ tens and _____ ones

4 29 _____ tens and _____ ones

Use a Number Construction Mat and Base-Ten Blocks or a Place Value Mat to build each addition problem. Say how many tens and ones are shown.

5 21 + 11

_____ tens and _____ ones = _____

6 24 + 15

_____ tens and _____ ones = _____

7 16 + 13

_____ tens and _____ ones = _____

8 17 + 12

_____ tens and _____ ones = _____

Practice

Regroup each number into tens and ones.

9 16 = _____ + _____

10 13 = _____ + _____

11 11 = _____ + _____

12 12 = _____ + _____

Regroup each number into tens and ones. Then find each sum.

13)
16 = 10 + 6
+ 13 = 10 + 3
_____
20 + 9 = ____

14)
22 = 20 + 2
+ 17 = 10 + 7
_____
____ + ____ = ____

15)
17 = ____ + ____
+ 12 = ____ + ____
_____
____ + ____ = ____

16)
11 = ____ + ____
+ 24 = ____ + ____
_____
____ + ____ = ____

17)
13 = ____ + ____
+ 25 = ____ + ____
_____
____ + ____ = ____

18)
19 = ____ + ____
+ 16 = ____ + ____
_____
____ + ____ = ____

19)
17 = ____ + ____
+ 11 = ____ + ____
_____
____ + ____ = ____

20)
27 = ____ + ____
+ 13 = ____ + ____
_____
____ + ____ = ____

Reflect

Using a Number Construction Mat and Base-Ten Blocks, draw a model of an addition problem that has a sum of 47.

Tens	Ones

Using Nice Numbers • Lesson 2

Week 2 — Using Nice Numbers

Lesson 3

Key Idea
Regrouping numbers into tens and ones is a strategy for adding two-digit numbers.

$$23 = 20 + 3$$
$$+\ 46 = 40 + 6$$
$$\overline{60 + 9 = 69}$$

Try This
Use a Number Construction Mat and Base-Ten Blocks or a Place Value Mat to build the following numbers. Show how many tens and ones you made.

① 62 _____ tens and _____ ones

② 74 _____ tens and _____ ones

③ 48 _____ tens and _____ ones

④ 95 _____ tens and _____ ones

Rewrite each number as tens and ones.

⑤ 56 = _____ + _____

⑥ 92 = _____ + _____

⑦ 64 = _____ + _____

⑧ 18 = _____ + _____

Practice
Use the Place Value Mat to build each addition problem. Show how many tens and ones are in the sum.

⑨ 28 + 31

_____ tens and _____ ones = _____

⑩ 52 + 15

_____ tens and _____ ones = _____

⑪ 77 + 12

_____ tens and _____ ones = _____

⑫ 64 + 22

_____ tens and _____ ones = _____

**Regroup each number into tens and ones.
Then find each sum.**

⑬ 54 = ____ + ____
  + 23 = ____ + ____
         ____ + ____ = ____

⑭ 18 = ____ + ____
  + 71 = ____ + ____
         ____ + ____ = ____

⑮ 66 = ____ + ____
  + 32 = ____ + ____
         ____ + ____ = ____

⑯ 84 = ____ + ____
  + 13 = ____ + ____
         ____ + ____ = ____

Regroup to find each sum.

⑰   62
   + 27

⑱   43
   + 25

⑲   51
   + 18

⑳   23
   + 36

㉑   48
   + 10

㉒   74
   + 26

Reflect
Jackie thinks it is easier to add 50 + 40 + 6 + 3 than 56 + 43. Do you agree with her? Why or why not?

_____

_____

Using Nice Numbers • Lesson 3

Week 2 — Using Nice Numbers

Lesson 4

Key Idea

"Leaps of 10" is an addition strategy that uses a number line to find the sum.

Use the following steps to help you find sums.
Step 1 Locate the first number on a number line.
Step 2 Break the second number into groups of tens and ones.
Step 3 Make "leaps of 10" and "hops of 1."

Try This
Write how many leaps and hops you make for each number.

① 34   Leaps _____ Hops _____

② 7   Leaps _____ Hops _____

③ 49   Leaps _____ Hops _____

④ 25   Leaps _____ Hops _____

⑤ 36   Leaps _____ Hops _____

⑥ 12   Leaps _____ Hops _____

Practice
Break down the following numbers into tens and ones.

⑦ 52 = _____ tens, _____ ones = 10 + 10 + 10 + 10 + 10 + 1 + 1

⑧ 44 = _____ tens, _____ ones = _____

⑨ 39 = _____ tens, _____ ones = _____

⑩ 21 = _____ tens, _____ ones = _____

Make leaps and hops on the number line to find each sum. Complete the addition sentences.

⑪ 20 + 10 = _____

⑫ 47 + 10 = _____

⑬ 54 + 20 = 54 + _____ + _____ = _____

⑭ 16 + 41 = 16 + _____ + _____ + _____ + _____ + _____ = _____

⑮ 62 + 34 = _____ = _____

⑯ 29 + 22 = _____ = _____

Reflect

Monica says that when you add 42 + 23, you need to make two moves on the number line to find the sum. Is she correct? Explain.

_____
_____
_____
_____

What is the sum of 42 + 23? _____

Using Nice Numbers

Lesson 5 Review

This week you explored addition strategies. You used "nice numbers," regrouping, and "leaps of 10" to find the sums.

Lesson 1 Circle the numbers below that are considered "nice numbers."

① 20   31   55   7   94   60   73   68   3   10   19

Write a new addition sentence by sharing the values. Make sure the new sentence includes a "nice number."

② 17 + 8

New addition sentence:
_____

③ 22 + 17

New addition sentence:
_____

Lesson 2 Use a Number Construction Mat and Base-Ten Blocks or a Place Value Mat to build each addition problem. Write how many tens and ones are shown.

④ 41 + 33   ____ tens and ____ ones = ____

⑤ 28 + 31   ____ tens and ____ ones = ____

⑥ 17 + 22   ____ tens and ____ ones = ____

⑦ 63 + 25   ____ tens and ____ ones = ____

Reflect

Explain the sharing you did to write the new addition sentence in Problem 3.

_____

_____

Lesson 3 Regroup each number into tens and ones. Then find each sum.

8   43 = 40 + 3
     + 32 = 30 + 2
     _____

9   15 = _____
     + 63 = _____
     _____

10   13 = _____
     + 25 = _____
     _____

11   51 = _____
     + 46 = _____
     _____

Lesson 4 Make leaps and hops on the number line to find each sum.

12   23 + 43 = 23 + 10 + 10 + 10 + 10 + 1 + 1 + 1

        = _____

13   51 + 34 = 51 + 10 + 10 + 10 + 1 + 1 + 1 + 1 = _____

Reflect
Draw a number line to illustrate the following addition problem using "leaps of 10" and "hops of 1." Explain your moves.

               56 + 23

Week 3 — Using Grouping

Lesson 1

Key Idea
When adding, the order of the numbers does not matter.

1 2 3 4 5 6 7 8 9 10

$6 + 4 = 10$

1 2 3 4 5 6 7 8 9 10

$4 + 6 = 10$

Try This
Find each sum.

1. $8 + 3 =$ _____ $3 + 8 =$ _____
2. $5 + 7 =$ _____ $7 + 5 =$ _____
3. $5 + 4 =$ _____ $4 + 5 =$ _____
4. $20 + 10 =$ _____ $10 + 20 =$ _____

Reorder the numbers in each addition problem to make "nice numbers." Then find the sum.

5. $6 + 3 + 4 = 6 + 4 + 3$

   _____ + _____ = _____

6. $9 + 5 + 1 = 9 + 1 + 5$

   _____ + _____ = _____

7. $13 + 7 + 4 =$ _____ + _____ + 4

   _____ + _____ = _____

22 Addition • Week 3

Reorder and group the addends to make "nice numbers" to help you find each sum.

8 $15 + 3 + 8 + 7 + 5$

$(15 + \underline{}) + (7 + \underline{}) + 8$

$\underline{} + \underline{} + \underline{} = \underline{}$

9 $4 + 9 + 16 + 11 + 7$

$(16 + \underline{}) + (11 + \underline{}) + \underline{}$

$\underline{} + \underline{} + \underline{} = \underline{}$

10 $21 + 15 + 5 + 7 + 9$

$(\underline{} + \underline{}) + (\underline{} + \underline{}) + \underline{}$

$\underline{} + \underline{} + \underline{} = \underline{}$

11 $8 + 13 + 7 + 12 + 5$

$(\underline{} + \underline{}) + (\underline{} + \underline{}) + \underline{}$

$\underline{} + \underline{} + \underline{} = \underline{}$

Reflect

Reorder and group the addends to make the problem easier to solve. Explain why this makes it easier to solve.

$18 + 4 + 16 + 10 + 2$

_____

_____

_____

Using Grouping • Lesson 1

Week 3 — Using Grouping
Lesson 2

Key Idea
Two strategies to make adding easier are the following:

1. Reorder Using Doubles Facts:
   $4 + 3 + 4 = (4 + 4) + 3 = 8 + 3 = 11$

2. Reorder Using "Nice Numbers":
   $2 + 5 + 8 = (8 + 2) + 5 = 10 + 5 = 15$

Try This
Write an addition sentence for each Dot Set Card.

1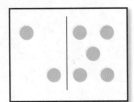

2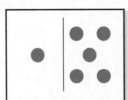

3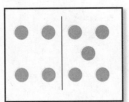

4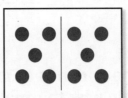

Use the Dot Set Cards to write an addition problem.
Find the sum. Circle the strategy you used.

5

____ + ____ + ____ + ____ = ____

Doubles Fact   or   "Nice Numbers"

6 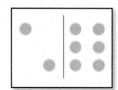 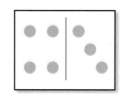   ___ + ___ + ___ + ___ = ___

Doubles Fact   or   "Nice Numbers"

Practice
Use the Dot Set Cards to write an addition problem. Find the sum.

7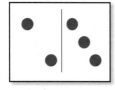

___ + ___ + ___ + ___ = ___

8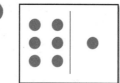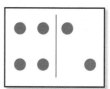

___ + ___ + ___ + ___ = ___

Use two Dot Set Cards to create an addition sentence that has each sum below. Draw the Dot Set Cards you used to create the sum, and then write the addition sentence below.

9 12 =

___ + ___ + ___ + ___ = 12

10 14 =

___ + ___ + ___ + ___ = 14

Reflect
Use Dot Set Cards to write two number sentences that have a sum of 15. Create one using the doubles-facts strategy and one using the "nice-numbers" strategy.

1. Doubles Fact: 15 = ___ + ___ + ___ + ___
2. Regrouping with "Nice Numbers":   15 = ___ + ___ + ___ + ___

Week 3 **Using Grouping**

Lesson 3

Key Idea

You can use more than one strategy when adding several numbers.

Doubles Facts	Near-Doubles Facts	"Nice Numbers"
4 + 7 + 4 =	9 + 4 + 5 =	6 + 3 + 4 =
(4 + 4) + 7 =	9 + (5 + 5 − 1) =	(6 + 4) + 3 =
8 + 7 =	9 + (10 − 1) =	10 + 3 =
15	18	13

Try This

Write the sums of both number sentences. Determine whether the calculator input is a correct way of finding it.

1 Original number sentence: 7 + 3 + 5 + 4 = ____

   Calculator input: 10 + 9 = ____

   Does it work?   ____

2 Original number sentence: 4 + 3 + 10 + 4 = ____

   Calculator input: 10 + 8 + 4 = ____

   Does it work?   ____

3 Original number sentence: 8 + 4 + 2 = ____

   Calculator input: 10 + 6 = ____

   Does it work?   ____

4 Original number sentence: 9 + 4 + 7 + 3 = ____

   Calculator input: 10 + 3 + 10 = ____

   Does it work?   ____

⑤ Original number sentence: 4 + 3 + 8 + 3 = _____

Calculator input: 8 + 8 + 2 = _____

Does it work? _____

⑥ Original number sentence: 3 + 8 + 3 + 2 = _____

Calculator input: 10 + 9 = _____

Does it work? _____

Practice
Find each sum. Write the strategy or strategies that you used.

⑦ 10 + 7 + 2 = _____

⑧ 7 + 7 + 2 + 5 = _____

⑨ 3 + 9 + 3 + 9 = _____

⑩ 4 + 6 + 3 + 6 = _____

Reflect
Write a number string of 4 numbers for each sum.

21 _____

18 _____

Using Grouping • Lesson 3

Week 3 — Using Grouping

Lesson 4

> **Key Idea**
> Watch out for numbers in the word problem that are not related to what the question is asking.

Try This
Solve each word problem using addition. Write the number sentence.

1. Kevyn went to a carnival three days in a row. On Friday he rode 4 rides. On Saturday he rode 6 rides. On Sunday he rode 4 rides. How many rides did Kevyn ride in all?

_____

2. Harold wanted to win tickets from the sports booth. He bought 3 tickets the first round, but he did not win. He bought 5 tickets the second round, but he did not win. Then he bought 7 tickets the third round and finally won. How many total tickets did Harold buy?

_____

3. Elizabeth sold 12 hot dogs during the first hour working at the food stand. During the second hour, she sold 11 hot dogs. During the third and fourth hours, she sold 8 and 10 hot dogs. How many total hot dogs did Elizabeth sell during her four-hour shift?

_____

Practice
Solve each word problem using addition. Write the number sentence.

④ A clown making balloon animals used 4 balloons to make a dog. He used 7 balloons to create a monkey. The giraffe he made required 5 balloons. He made a bird using 4 balloons. How many total balloons did the clown use to make these animals?

_____

⑤ Margaret started selling tickets for the cake booth at 1:00. She sold the first set in 8 minutes. It took her 5 minutes to sell the second set and 9 minutes to sell the third set. How long did it take for Margaret to sell the three sets of tickets?

_____

Reflect
Angie solved the following problem.

Tyrese bought fish for his fish tank three days in a row this week. On Monday he bought 7 fish. On Tuesday he bought 8 fish. On Wednesday he bought 5 fish. How many fish did Tyrese buy in all?

Angie wrote 3 + 7 + 8 + 5 = 23 fish. Is she correct? If not, what should she have written?

_____

_____

_____

_____

Using Grouping • Lesson 4

Week 3 — Using Grouping

Lesson 5 Review

This week you explored more addition strategies. You learned that the order of the addends does not affect the sum, so you can reorder the addends to use "nice numbers," doubles facts, and near-doubles facts.

Lesson 1 Answer the following questions.

① Does 8 + 3 = 3 + 8? _____   ② Does 7 + 5 = 5 + 7? _____

If yes, what is the sum? _____   If yes, what is the sum? _____

③ What do you notice about the two problems above?

_____

Lesson 2 Write two number strings for the sum. Use the doubles-facts strategy for one and the "nice-numbers" strategy for the other.

④ 15

Doubles Fact: _____

"Nice Number": _____

Reflect

Show how to reorder the addends to use doubles facts in the problem below. Find the sum. Then show how to reorder the addends to use "nice numbers." Find the sum.

7 + 8 + 3 + 7

_____

_____

30 Addition • Week 3

Lesson 3 Reorder the addends in the addition problem to make "nice numbers." Then find the sum.

5 6 + 9 + 17 + 14 + 3

(___ + ___) + (___ + ___) + ___

___ + ___ + ___ = ___

Lesson 4 Solve each word problem using addition. Write the number sentence.

6 Brittany and Kimberly played **Race for the Sum**. Their first round took 12 minutes. The second round took 16 minutes, and the third round was over in 8 minutes. The final round was finished in 10 minutes. How long did the girls play?

_____

7 John was in charge of setting up the carnival booths. There were three rows of booths to set up. The first row took 20 minutes, and the second row took 18 minutes. John needed 22 minutes for the third row. How long did John take to put up all the booths?

_____

Reflect
Explain why addends were reordered in the problem below. Is the sum correct? If not, correct the mistake.

22 + 9 + 13 + 8 + 7   (22 + 8) + (13 + 7) + 9

20   +   20   + 9 = 49

_____

_____

_____

Using Grouping • Lesson 5 Review

Week 4 — Using Partial Sums

Lesson 1

Key Idea
Models and pictures can help with renaming when adding two-digit numbers.

Try This
Circle each set of 10 unit blocks below. Redraw the value given using rods and unit blocks. Write the value.

Practice
Use Base-Ten Blocks to create a model for each number. Rename ones to tens as needed. Find the sum.

5) 62 + 29

6) 45 + 37

7) 39 + 14

8) 77 + 18

Draw a picture using Base-Ten Blocks for each problem. Find the sum.

9) 49 + 47

10) 67 + 18

Reflect
Look at the following problem. Is the answer correct? Explain.

```
  59
+ 46
----
 915
```

Week 4 — Using Partial Sums

Lesson 2

> **Key Idea**
> Models and pictures can help you add three-digit numbers with renaming.

Try This

Circle each set of 10 unit blocks below. Circle each set of 10 rods below. Redraw the value given using flats, rods, and unit blocks. Write the value.

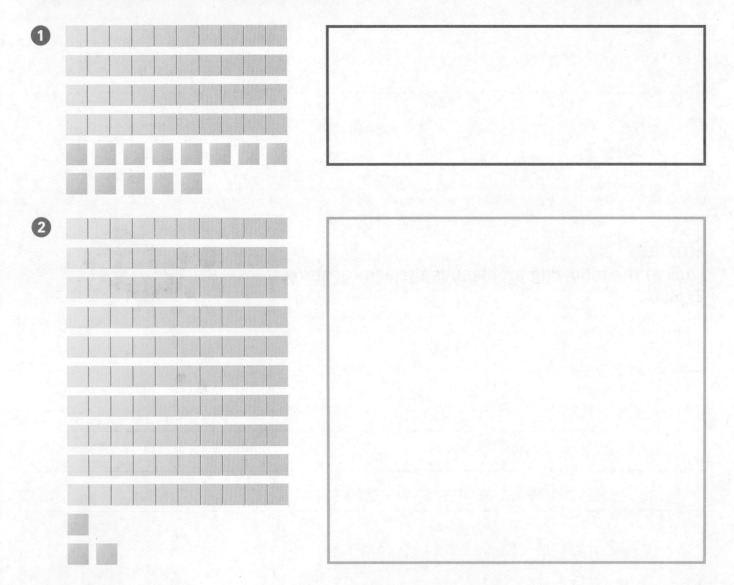

34 Addition • Week 4

Practice

Use Base-Ten Blocks to create a model for each number. Rename ones to tens, and tens to hundreds as needed. Find the sum.

3) 137
   + 325

4) 275
   + 142

5) 435
   + 236

6) 378
   + 163

Find the sum.

7) 639
   + 145

8) 434
   + 567

Reflect
Explain the renaming that you did in Problem 8.

_____
_____
_____
_____

Using Partial Sums • Lesson 2

Week 4 — Using Partial Sums

Lesson 3

> **Key Idea**
> Find the sum of the tens digits and the sum of the ones digits, and then add them. This strategy is called *partial sums*.
>
> ```
>   27
> +46
>   60   (20 + 40) Add tens.
> +13   (7 + 6) Add ones.
>   73
> ```

Try This

Write each number as the sum of tens and ones values.

① 46 = ____ + ____

② 71 = ____ + ____

③ 35 = ____ + ____

④ 62 = ____ + ____

Fill in the missing values using the partial-sums strategy.

⑤
```
    35
+   17
```
____ (30 + 10)

+____ (5 + 7)

⑥
```
    64
+   21
```
____ (60 + 20)

+____ (4 + 1)

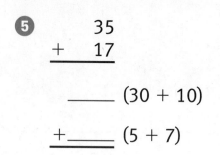

⑦
```
    32
+   56
```
80 (____ + ____)

+   8 (____ + ____)

⑧
```
    49
+   15
```
50 (____ + ____)

+ 14 (____ + ____)

36 Addition • Week 4

Practice
Fill in the missing values using the partial-sums strategy.

9   26
   + 43

   ____ (____ + ____)
   +____ (____ + ____)

10   13
   + 39

   ____ (____ + ____)
   +____ (____ + ____)

Find each sum using the partial-sums strategy.

11   70
   + 22

12   54
   + 21

13   28
   + 63

14   36
   + 43

Reflect
Use the partial-sums strategy to find the sum. Explain each of your steps.

   54
+ 38

_____

_____

_____

Using Partial Sums • Lesson 3

Week 4 — Lesson 4
Using Partial Sums

Key Idea
The partial-sums strategy can also be used with three-digit numbers.

```
    314
  + 158
    400    (300 + 100) Add hundreds.
     60    (10 + 50) Add tens.
  +  12    (4 + 8) Add ones.
    472
```

Try This
Write each number as the sum of hundreds, tens, and ones values.

① 182 = ____ + ____ + ____

② 719 = ____ + ____ + ____

③ 304 = ____ + ____

④ 520 = ____ + ____

Fill in the missing values using the partial-sums strategy.

⑤
```
    243
  + 151
```
____ (200 + 100)

____ (40 + 50)

+____ (3 + 1)

⑥
```
    126
  + 315
```
____ (100 + 300)

____ (20 + 10)

+____ (6 + 5)

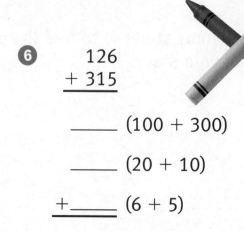

38   Addition • Week 4

Practice

Fill in the missing values using the partial-sums strategy.

7)
```
   436
+ 261
```
_____ (_____ + _____)
_____ (_____ + _____)
+_____ (_____ + _____)

8)
```
   753
+ 138
```
_____ (_____ + _____)
_____ (_____ + _____)
+_____ (_____ + _____)

Find each sum using the partial-sums strategy.

9)
```
   642
+ 153
```

10)
```
   320
+ 246
```

11)
```
   343
+ 111
```

12)
```
   227
+ 516
```

Reflect

Use the partial-sums strategy to find the sum. Explain each of your steps.

```
   743
+ 352
```

_____

_____

_____

Using Partial Sums • Lesson 4

Using Partial Sums

Lesson 5 Review

This week you explored more addition strategies. You used partial sums and Base-Ten Blocks with renaming to find sums.

Lesson 1 Use the partial-sums strategy to find each sum.

①   38
    + 55

    _____ (30 + 50)
   +_____ (8 + 5)

②   138
    + 312

    _____ (100 + 300)
    _____ (30 + 10)
   +_____ (8 + 2)

Lesson 2 Use the partial-sums strategy to find each sum.

③   37
    + 44

    _____ (_____ + _____)
   +_____ (_____ + _____)

④   536
    + 368

    _____ (_____ + _____)
    _____ (_____ + _____)
   +_____ (_____ + _____)

Reflect
Name the partial sums for the problem below.
Find the sum.

   556
+ 245

Lesson 3 Use Base-Ten Blocks to create a model for each number. Rename ones to tens and tens to hundreds as needed. Draw your model. Find the sum.

5)
```
   64
 + 27
```

6)
```
   34
 + 19
```

7)
```
   78
 + 17
```

Lesson 4 ## Practice
Use Base-Ten Blocks to create a model for each number. Rename ones to tens and tens to hundreds as needed. Find the sum.

8)
```
   177
 + 286
```

9)
```
   454
 + 277
```

10)
```
   194
 + 268
```

Reflect
Which problem below does not require any renaming to find the sum? Explain.

11)
```
   31
 + 55
```

12)
```
   268
 + 513
```

13)
```
   344
 + 292
```

Using Partial Sums • Lesson 5 Review

Week 1 — Using Doubles and Near Doubles

Practice

Use doubles facts to find the sums of each near-doubles fact.

1. If 20 + 20 = _____, then

    20 + 21 = _____.

2. If 55 + 55 = _____, then

    55 + 56 = _____.

Use doubles facts to find the missing addend.

3. 12 + _____ = 23

4. 14 + _____ = 29

Use the Number Construction Mat and Base-Ten Blocks to create a model of the addition problems. Find each sum.

5. 14 + 10 = _____

6. 46 + 10 = _____

Using a 100 Chart, find each sum.

7. 47 + 9 = _____

8. 19 + 9 = _____

Use +10 combinations to help you decide if the missing addend is 9 or 11. Circle your choice.

9. 18 + (9 or 11) = 29

10. 22 + (9 or 11) = 31

Using doubles facts, near-doubles facts, +10 combinations, +9 combinations, or +11 combinations, make an addition problem for the following sum. Write the strategy used.

11. 39 = _____ + _____   _____

42 Addition • Week 1 Practice

Week 2 — Using Nice Numbers

Practice

Circle the numbers below that are considered "nice numbers."

❶ 18   40   96   11   29   70   35   10   24   33   64

Write a new addition sentence by sharing the values. Be sure that a "nice number" is in the new sentence.

❷ 56 + 41

New addition sentence

_____

❸ 17 + 62

New addition sentence

_____

Use a Number Construction Mat and Base-Ten Blocks or a Place Value Mat to build each addition problem. Say how many tens and ones are modeled.

❹ 23 + 74

____ tens and ____ ones = ____

❺ 46 + 12

____ tens and ____ ones = ____

❻ 51 + 18

____ tens and ____ ones = ____

❼ 36 + 41

____ tens and ____ ones = ____

Regroup each number into tens and ones. Then find each sum.

❽ 42 = 40 + 2
   + 16 = 10 + 6
   _____
   _____

❾ 26 = _____
   + 61 = _____
   _____
   _____

Addition • Week 2 Practice   **43**

Using Grouping

Practice

Answer the following questions.

① Does 6 + 4 = 4 + 6? _____   If yes, what is the sum? _____

② What do you notice about the problem above?
_____

Reorder the numbers in each addition problem to make "nice numbers." Then find the sum.

③ 3 + 6 + 7 = _____   ④ 9 + 8 + 2 + 1 = _____

Reorder the numbers in the addition problem to make "nice numbers." Then find the sum.

⑤ 7 + 5 + 13 + 4 + 5

(____ + ____) + (____ + ____) + ____

____ + ____ + ____ = ____

Solve the word problem using addition. Write the number sentence.

⑥ Travis sets aside time for reading every night. Four nights ago, Travis read 26 pages of a book. He read 31 pages three nights ago and 21 pages two nights ago. Finally he read 30 pages last night. How many pages did Travis read over the last four nights?

_____

Week 4 — Using Partial Sums

Practice

Use the partial-sums strategy to find each sum.

1)
```
  284
+ 685
```
____ (200 + 600)

____ (80 + 80)

+____ (4 + 5)

2)
```
  158
+ 421
```
500 (____ + ____)

70 (____ + ____)

+ 9 (____ + ____)

Use the partial-sums strategy to find each sum.

3)
```
  46
+ 13
```
____ (____ + ____)

+____ (____ + ____)

4)
```
  824
+ 115
```
____ (____ + ____)

____ (____ + ____)

+____ (____ + ____)

Use Base-Ten Blocks to create a model for each number. Rename ones to tens and tens to hundreds as needed. Draw your model. Find the sum.

5)
```
  26
+ 37
```

6)
```
  14
+ 38
```

7)
```
  629
+ 187
```

8)
```
  557
+ 385
```

Addition • Week 4 Practice   45

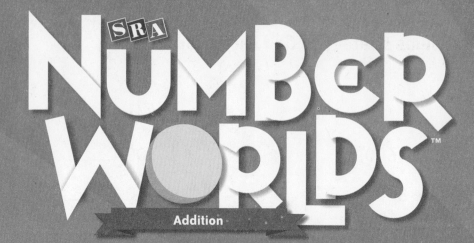

Unit 3 Workbook

SRAonline.com

Level E

R5313X.01

Unit 3 Workbook
Level E

SRA
NUMBER WORLDS
Addition

featuring Building Blocks Software

Author
Sharon Griffin
*Associate Professor of Education and
Adjunct Associate Professor of Psychology*
Clark University
Worcester, Massachusetts

Building Blocks Authors

Douglas H. Clements
*Professor of Early Childhood
and Mathematics Education*
University at Buffalo
State University of New York, New York

Julie Sarama
Associate Professor of Mathematics Education
University at Buffalo
State University of New York, New York

Contributing Writers
Sherry Booth, *Math Curriculum Developer,* Raleigh, North Carolina
Elizabeth Jimenez, *English Language Learner Consultant,* Pomona, California

Program Reviewers

Jean Delwiche
Almaden Country School
San Jose, California

Cheryl Glorioso
Santa Ana Unified School District
Santa Ana, California

Sharon LaPoint
School District of Indian River County
Vero Beach, Florida

Leigh Lidrbauch
Pasadena Independent School District
Pasadena, Texas

Dave Maresh
Morongo Unified School District
Yucca Valley, California

Mary Mayberry
Mon Valley Education Consortium, AIU 3
Clairton, Pennsylvania

Lauren Parente
Mountain Lakes School District
Mountain Lakes, New Jersey

Juan Regalado
Houston Independent School District
Houston, Texas

M. Kate Thiry
Dublin City School District
Dublin, Ohio

Susan C. Vohrer
Baltimore County Public Schools
Baltimore, Maryland

SRAonline.com

Copyright © 2007 SRA/McGraw-Hill.

All rights reserved. Except as permitted under the United States Copyright Act, no part of this publication may be reproduced or distributed in any form or by any means, or stored in a database or retrieval system, without the prior written permission of the publisher, unless otherwise indicated.

Printed in the United States of America.

Send all inquiries to:
SRA/McGraw-Hill
4400 Easton Commons
Columbus, OH 43219-6188

R5313X.01

9 WCE 12 11 10

Photo Credits

3–39 ©PhotoDisc/Getty Images, Inc.

The McGraw·Hill Companies

Contents

Addition

Week 1 Using Doubles and Near Doubles **2**

Week 2 Using Nice Numbers **12**

Week 3 Using Grouping **22**

Week 4 Using Partial Sums **32**

Week 1 Practice **42**

Week 2 Practice **43**

Week 3 Practice **44**

Week 4 Practice **45**

Week 1 — Using Doubles and Near Doubles

Lesson 1

Key Idea

2	+	2	=	4	**Doubles Fact**
2	+	3	=	2 + (2 + 1)	**Near-Doubles Fact**
			=	4 + 1 = 5	

Try This

Find the sum of each doubles fact.

① 2 + 2 = _____

② 5 + 5 = _____

③ 4 + 4 = _____

④ 1 + 1 = _____

Use a doubles fact to find the sum of each near-doubles fact.

⑤ 2 + 3 = 2 + (2 + 1)
   ∨
   _____ + 1 = _____

⑥ 5 + 6 = 5 + (5 + 1)
   ∨
   _____ + 1 = _____

⑦ 3 + 4 = 3 + (3 + 1)
   ∨
   _____ + 1 = _____

⑧ 4 + 5 = 4 + (4 + 1)
   ∨
   _____ + 1 = _____

2 Addition • Week 1

Practice
Use the sum of each doubles fact to find the sum of each near-doubles fact.

9 If 3 + 3 = _____, then

3 + 4 = _____.

10 If 5 + 5 = _____, then

5 + 6 = _____.

11 If 7 + 7 = _____, then

7 + 8 = _____.

12 If 8 + 8 = _____, then

8 + 9 = _____.

13 If 5 + 5 = _____, then

5 + 4 = _____.

14 If 2 + 2 = _____, then

1 + 2 = _____.

15 If 7 + 7 = _____, then

6 + 7 = _____.

16 If 4 + 4 = _____, then

3 + 4 = _____.

Find each sum.

17 3 + 4 = _____

18 5 + 6 = _____

19 10 + 9 = _____

20 7 + 6 = _____

21 8 + 7 = _____

22 5 + 4 = _____

Reflect
Given the problem 4 + 5, show how each of the doubles facts below can be used to find the sum.

4 + 4   or   5 + 5

_____ or _____

Using Doubles and Near Doubles • Lesson 1

Week 1: Using Doubles and Near Doubles

Lesson 2

Key Idea

When 10 is added to a number, the digit in the tens place increases by 1. You can use a 99 Chart to add by circling the starting number, using an arrow to show a jump of 10, and shading in the sum.

12 + 10 = 22									
21	**22**	23	24	25	26	27	28	29	30
11	(12)	13	14	15	16	17	18	19	20

Try This

Use the Number Construction Mat and Base-Ten Blocks to create a model of the addition problems. Find each sum.

① 20 + 10 = _____

② 50 + 10 = _____

③ 24 + 10 = _____

④ 38 + 10 = _____

Use a 99 Chart to find each sum.

⑤ 80 + 10 = _____

⑥ 60 + 10 = _____

⑦ 51 + 10 = _____

⑧ 72 + 10 = _____

⑨ In what direction do you move on a 99 Chart when 10 is added to a number?

_____

4 Addition • Week 1

Practice

Use the Number Construction Mat and Base-Ten Blocks to create a model of the addition problems. Find each sum.

⑩ 13 + 10 = _____

⑪ 49 + 10 = _____

⑫ 36 + 10 = _____

⑬ 65 + 10 = _____

Use a 99 Chart to find each missing addend.

⑭ _____ + 10 = 40

⑮ _____ + 10 = 61

⑯ _____ + 10 = 66

⑰ _____ + 10 = 94

Use the Number Construction Mat and Base-Ten Blocks to create a model of the addition problems. Find each missing addend.

⑱ _____ + 10 = 29

⑲ _____ + 10 = 77

⑳ _____ + 10 = 37

㉑ _____ + 10 = 55

Reflect

What pattern do you notice when 10 is added to any number?
_____

Week 1 — Using Doubles and Near Doubles

Lesson 3

Key Idea

On a 99 Chart:
Add 10 .... Move UP ↑
Add 1 .... Move RIGHT →
Subtract 1 ...... Move LEFT ←

Try This

Use a 99 Chart to find each sum or difference. Circle the correct direction arrow on the 99 Chart to find the sum or difference.

1. 51 + 10 = _____   ↑   →   ←

2. 51 + 1 = _____   ↑   →   ←

3. 51 − 1 = _____   ↑   →   ←

4. 47 + 10 = _____   ↑   →   ←

5. 47 + 1 = _____   ↑   →   ←

6. 47 − 1 = _____   ↑   →   ←

Use a 99 Chart to model each problem. Find each sum. Circle the correct direction arrow or arrows on a 99 Chart to find the sum.

7. 58 + 10 = _____   ↑   →   ←

8. 63 + 11 =
   63 + (10 + 1) = _____   ↑   →   ←

9. 15 + 9 =
   15 + (10 − 1) = _____   ↑   →   ←

6 Addition • Week 1

Practice

Use a 99 Chart to model each problem. Find each sum and circle the direction or directions that you moved on the 99 Chart.

⑩ 16 + 10 = _____
    Start at 16 and move **up or down** 1 block.

⑪ 16 + 9 = _____
    Start at 16 and move **up or down** 1 block and **left or right** 1 block.

⑫ 16 + 11 = _____
    Start at 16 and move **up or down** 1 block and **left or right** 1 block.

⑬ 34 + 10 = _____
    Start at 34 and move **up or down** 1 block.

⑭ 34 + 9 = _____
    Start at 34 and move **up or down** 1 block and **left or right** 1 block.

⑮ 34 + 11 = _____
    Start at 34 and move **up or down** 1 block and **left or right** 1 block.

Reflect

What do you do differently using a 99 Chart when adding 9 rather than 11 to a number?

_____

_____

_____

Week 1 — Using Doubles and Near Doubles

Lesson 4

Key Idea

$10 + 10 = 20$

$10 + 9 = 19$
$10 + (10 - 1) = 19$
$20 - 1 = 19$

$10 + 11 = 21$
$10 + (10 + 1) = 21$
$20 + 1 = 21$

Try This
Find the sum of each doubles fact.

① $3 + 3 = $ _____

② $30 + 30 = $ _____

③ $4 + 4 = $ _____

④ $40 + 40 = $ _____

Use a doubles fact to find the sum of each near-doubles fact.

⑤ $8 + 9 = 8 + (8 + 1)$
_____ $+ 1 = $ _____

⑥ $8 + 7 = 8 + (8 - 1)$
_____ $- 1 = $ _____

⑦ $20 + 21 = 20 + (20 + 1)$
_____ $+ 1 = $ _____

⑧ $20 + 19 = 20 + (20 - 1)$
_____ $- 1 = $ _____

⑨ $40 + 41 = 40 + (40 + 1)$
_____ $+ 1 = $ _____

⑩ $40 + 39 = 40 + (40 - 1)$
_____ $- 1 = $ _____

⑪ $25 + 26 = 25 + (25 + 1)$
_____ $+ 1 = $ _____

⑫ $25 + 24 = 25 + (25 - 1)$
_____ $- 1 = $ _____

Practice

Use the sum of each doubles fact to find the sum of each near-doubles fact.

13 If 30 + 30 = ____, then

30 + 31 = ____.

14 If 15 + 15 = ____, then

15 + 14 = ____.

15 If 30 + 30 = ____, then

30 + 29 = ____.

16 If 15 + 15 = ____, then

15 + 16 = ____.

17 If 50 + 50 = ____, then

50 + 49 = ____.

18 If 45 + 45 = ____, then

45 + 46 = ____.

Use doubles facts to find each missing addend.

19 7 + ____ = 15

20 6 + ____ = 13

21 30 + ____ = 61

22 25 + ____ = 49

Reflect

Explain how using a doubles fact helps you find the sum of a near-doubles fact.

_____

_____

_____

Using Doubles and Near Doubles • Lesson 4

Week 1

Using Doubles and Near Doubles

Lesson 5 Review

This week you explored addition strategies. You discovered that doubles facts are helpful tools when working with near-doubles facts.

Lesson 1 Use a doubles fact to find the sum of each near-doubles fact.

1. If 40 + 40 = _____, then 40 + 41 = _____.

2. If 25 + 25 = _____, then 25 + 24 = _____.

3. If 35 + 35 = _____, then 35 + 36 = _____.

4. If 50 + 50 = _____, then 49 + 50 = _____.

Lesson 2 Use near-doubles facts to find each missing addend.

5. 7 + _____ = 13

6. 9 + _____ = 17

7. 30 + _____ = 61

8. 15 + _____ = 29

Use the Number Construction Mat and Base-Ten Blocks to create a model of the addition problems. Find each sum.

9. 16 + 10 = _____

10. 10 + 52 = _____

11. 43 + 10 = _____

12. 88 + 10 = _____

Reflect

Explain how to use a 99 Chart to find the sum of 38 + 9.

_____

_____

10 Addition • Week 1

Lesson 3 Using a 99 Chart, find each sum.

⑬ 54 + 9 = _____   ⑭ 87 + 9 = _____

⑮ 38 + 11 = _____   ⑯ 19 + 11 = _____

⑰ 22 + 9 = _____   ⑱ 61 + 9 = _____

Use +10 combinations to help you decide if the missing addend is 9 or 11. Circle your choice.

⑲ 15 + (9 or 11) = 26   ⑳ 55 + (9 or 11) = 66

㉑ 34 + (9 or 11) = 43   ㉒ 46 + (9 or 11) = 57

㉓ 79 + (9 or 11) = 88   ㉔ 58 + (9 or 11) = 67

Lesson 4 Using doubles facts, near-doubles facts, +10 combinations, +9 combinations, or +11 combinations, make an addition problem for the following sums. Use a different strategy for each. Write the strategy used on the line provided.

㉕ 54 = _____ + _____ _____

㉖ 48 = _____ + _____ _____

㉗ 66 = _____ + _____ _____

㉘ 39 = _____ + _____ _____

Reflect
Which strategies do you like to use best? Why?

_____

_____

_____

Week 2 — Using Nice Numbers

Lesson 1

Key Idea

6 + 4 = 10

18 + 2 = 20

A **"nice number"** is a number that **ends in a 0.**

Try This
On the number line, circle all the "nice numbers."

1

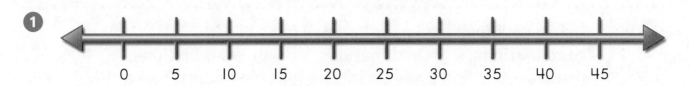

0   5   10   15   20   25   30   35   40   45

Using the number line above, find the given number's closest "nice number."

2 23 _____   **3** 18 _____   **4** 41 _____

5 28 _____   **6** 37 _____   **7** 32 _____

Practice
Make a "nice number" by drawing the missing boxes.
Find the missing addend.

8

7 + ____ = 10

9

____ + 19 = 20

Practice

Make a "nice number" by writing the missing addend.

10 ____ + 16 = 20

11 ____ + 5 = 10

Use the Number Construction Mat and Base-Ten Blocks to show a model of the following problems. Write an addition sentence using the "nice number" that is described.

12 19 + 3   Share 1 from 3 and add it to 19.

New addition sentence: _____

13 28 + 6   Share 2 from 6 and add it to 28.

New addition sentence: _____

Write a new addition sentence by sharing the values. Make sure that a "nice number" is in the new sentence.

14 13 + 6

New addition sentence: _____

15 34 + 11

New addition sentence: _____

Reflect

Give three addition problems that have a sum of 24. Then underline the number in each sentence that is closer to a nice number.

_____

_____

_____

Using Nice Numbers • Lesson 1

Week 2 — Using Nice Numbers

Lesson 2

Key Idea

You can regroup two-digit numbers into "nice numbers" to make them easier to work with.

24 = 20 + 4

Tens	Ones
(2 rods shown) Its value is 20.	(4 units shown) Its value is 4.

Try This

Use a Number Construction Mat and Base-Ten Blocks or a Place Value Mat to build each number. Say how many tens and ones are shown.

1. 24 _____ tens and _____ ones
2. 12 _____ tens and _____ ones
3. 35 _____ tens and _____ ones
4. 29 _____ tens and _____ ones

Use a Number Construction Mat and Base-Ten Blocks or a Place Value Mat to build each addition problem. Say how many tens and ones are shown.

5. 21 + 11
   _____ tens and _____ ones = _____

6. 24 + 15
   _____ tens and _____ ones = _____

7. 16 + 13
   _____ tens and _____ ones = _____

8. 17 + 12
   _____ tens and _____ ones = _____

Practice

Regroup each number into tens and ones.

9. 16 = _____ + _____
10. 13 = _____ + _____
11. 11 = _____ + _____
12. 12 = _____ + _____

Regroup each number into tens and ones.
Then find each sum.

⑬ 16 = 10 + 6
  + 13 = 10 + 3
        20 + 9 = ____

⑭ 22 = 20 + 2
  + 17 = 10 + 7
        ____ + ____ = ____

⑮ 17 = ____ + ____
  + 12 = ____ + ____
        ____ + ____ = ____

⑯ 11 = ____ + ____
  + 24 = ____ + ____
        ____ + ____ = ____

⑰ 13 = ____ + ____
  + 25 = ____ + ____
        ____ + ____ = ____

⑱ 19 = ____ + ____
  + 16 = ____ + ____
        ____ + ____ = ____

⑲ 17 = ____ + ____
  + 11 = ____ + ____
        ____ + ____ = ____

⑳ 27 = ____ + ____
  + 13 = ____ + ____
        ____ + ____ = ____

Reflect

Using a Number Construction Mat and Base-Ten Blocks, draw a model of an addition problem that has a sum of 47.

Tens	Ones

Using Nice Numbers • Lesson 2   15

Week 2 — Using Nice Numbers

Lesson 3

Key Idea

Regrouping numbers into tens and ones is a strategy for adding two-digit numbers.

$$23 = 20 + 3$$
$$+ 46 = 40 + 6$$
$$ 60 + 9 = 69$$

Try This

Use a Number Construction Mat and Base-Ten Blocks or a Place Value Mat to build the following numbers. Show how many tens and ones you made.

1 62 _____ tens and _____ ones

2 74 _____ tens and _____ ones

3 48 _____ tens and _____ ones

4 95 _____ tens and _____ ones

Rewrite each number as tens and ones.

5 56 = _____ + _____

6 92 = _____ + _____

7 64 = _____ + _____

8 18 = _____ + _____

Practice

Use the Place Value Mat to build each addition problem. Show how many tens and ones are in the sum.

9 28 + 31

_____ tens and _____ ones = _____

10 52 + 15

_____ tens and _____ ones = _____

11 77 + 12

_____ tens and _____ ones = _____

12 64 + 22

_____ tens and _____ ones = _____

16 Addition • Week 2

Regroup each number into tens and ones. Then find each sum.

13) 54 = _____ + _____
   + 23 = _____ + _____
   _____
         _____ + _____ = _____

14) 18 = _____ + _____
   + 71 = _____ + _____
   _____
         _____ + _____ = _____

15) 66 = _____ + _____
   + 32 = _____ + _____
   _____
         _____ + _____ = _____

16) 84 = _____ + _____
   + 13 = _____ + _____
   _____
         _____ + _____ = _____

Regroup to find each sum.

17) 62
   + 27

18) 43
   + 25

19) 51
   + 18

20) 23
   + 36

21) 48
   + 10

22) 74
   + 26

Reflect
Jackie thinks it is easier to add 50 + 40 + 6 + 3 than 56 + 43. Do you agree with her? Why or why not?

_____

_____

Using Nice Numbers • Lesson 3

Week 2 — Using Nice Numbers

Lesson 4

Key Idea

"Leaps of 10" is an addition strategy that uses a number line to find the sum.

Use the following steps to help you find sums.
Step 1 Locate the first number on a number line.
Step 2 Break the second number into groups of tens and ones.
Step 3 Make "leaps of 10" and "hops of 1."

Try This
Write how many leaps and hops you make for each number.

❶ 34   Leaps _____ Hops _____

❷ 7   Leaps _____ Hops _____

❸ 49   Leaps _____ Hops _____

❹ 25   Leaps _____ Hops _____

❺ 36   Leaps _____ Hops _____

❻ 12   Leaps _____ Hops _____

Practice
Break down the following numbers into tens and ones.

❼ 52 = _____ tens, _____ ones = 10 + 10 + 10 + 10 + 10 + 1 + 1

❽ 44 = _____ tens, _____ ones = _____

❾ 39 = _____ tens, _____ ones = _____

❿ 21 = _____ tens, _____ ones = _____

Make leaps and hops on the number line to find each sum. Complete the addition sentences.

⑪ 20 + 10 = _____

⑫ 47 + 10 = _____

⑬ 54 + 20 = 54 + _____ + _____ = _____

⑭ 16 + 41 = 16 + _____ + _____ + _____ + _____ + _____ = _____

⑮ 62 + 34 = _____ = _____

⑯ 29 + 22 = _____ = _____

Reflect

Monica says that when you add 42 + 23, you need to make two moves on the number line to find the sum. Is she correct? Explain.

_____

_____

_____

_____

What is the sum of 42 + 23? _____

Using Nice Numbers • Lesson 4

Week 2 — Using Nice Numbers

Lesson 5 Review

This week you explored addition strategies. You used "nice numbers," regrouping, and "leaps of 10" to find the sums.

Lesson 1 Circle the numbers below that are considered "nice numbers."

① 20   31   55   7   94   60   73   68   3   10   19

Write a new addition sentence by sharing the values. Make sure the new sentence includes a "nice number."

② 17 + 8

New addition sentence: _____

③ 22 + 17

New addition sentence: _____

Lesson 2 Use a Number Construction Mat and Base-Ten Blocks or a Place Value Mat to build each addition problem. Write how many tens and ones are shown.

④ 41 + 33   ____ tens and ____ ones = ____

⑤ 28 + 31   ____ tens and ____ ones = ____

⑥ 17 + 22   ____ tens and ____ ones = ____

⑦ 63 + 25   ____ tens and ____ ones = ____

Reflect

Explain the sharing you did to write the new addition sentence in Problem 3.

_____

_____

Lesson 3 Regroup each number into tens and ones. Then find each sum.

⑧ $\quad 43 = 40 + 3$
$\quad\;+\, 32 = 30 + 2$
$\quad$ _____

⑨ $\quad 15 = $ _____
$\quad\;+\, 63 = $ _____
$\quad$ _____

⑩ $\quad 13 = $ _____
$\quad\;+\, 25 = $ _____
$\quad$ _____

⑪ $\quad 51 = $ _____
$\quad\;+\, 46 = $ _____
$\quad$ _____

Lesson 4 Make leaps and hops on the number line to find each sum.

⑫ $23 + 43 = 23 + 10 + 10 + 10 + 10 + 1 + 1 + 1$
$\qquad\quad = $ _____

⑬ $51 + 34 = 51 + 10 + 10 + 10 + 1 + 1 + 1 + 1 = $ _____

Reflect

Draw a number line to illustrate the following addition problem using "leaps of 10" and "hops of 1." Explain your moves.

$$56 + 23$$

Week 3 — Using Grouping

Lesson 1

Key Idea
When adding, the order of the numbers does not matter.

$6 + 4 = 10$ $4 + 6 = 10$

Try This
Find each sum.

1) $8 + 3 = $ _____ $3 + 8 = $ _____

2) $5 + 7 = $ _____ $7 + 5 = $ _____

3) $5 + 4 = $ _____ $4 + 5 = $ _____

4) $20 + 10 = $ _____ $10 + 20 = $ _____

Reorder the numbers in each addition problem to make "nice numbers." Then find the sum.

5) $6 + 3 + 4 = 6 + 4 + 3$

_____ + _____ = _____

6) $9 + 5 + 1 = 9 + 1 + 5$

_____ + _____ = _____

7) $13 + 7 + 4 = $ _____ + _____ + 4

_____ + _____ = _____

22 Addition • Week 3

Reorder and group the addends to make "nice numbers" to help you find each sum.

⑧ 15 + 3 + 8 + 7 + 5

(15 + ___) + (7 + ___) + 8

___ + ___ + ___ = ___

⑨ 4 + 9 + 16 + 11 + 7

(16 + ___) + (11 + ___) + ___

___ + ___ + ___ = ___

⑩ 21 + 15 + 5 + 7 + 9

(___ + ___) + (___ + ___) + ___

___ + ___ + ___ = ___

⑪ 8 + 13 + 7 + 12 + 5

(___ + ___) + (___ + ___) + ___

___ + ___ + ___ = ___

Reflect

Reorder and group the addends to make the problem easier to solve. Explain why this makes it easier to solve.

18 + 4 + 16 + 10 + 2

_____

_____

_____

_____

Week 3 — Using Grouping

Lesson 2

Key Idea
Two strategies to make adding easier are the following:

1. **Reorder Using Doubles Facts:**
   $4 + 3 + 4 = (4 + 4) + 3 = 8 + 3 = 11$

2. **Reorder Using "Nice Numbers":**
   $2 + 5 + 8 = (8 + 2) + 5 = 10 + 5 = 15$

Try This
Write an addition sentence for each Dot Set Card.

1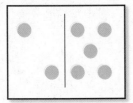

2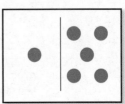

3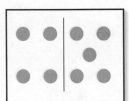

4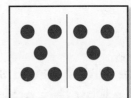

Use the Dot Set Cards to write an addition problem.
Find the sum. Circle the strategy you used.

5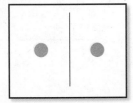

____ + ____ + ____ + ____ = ____

Doubles Fact   or   "Nice Numbers"

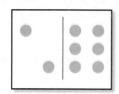 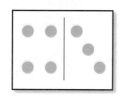 ___ + ___ + ___ + ___ = ___

Doubles Fact or "Nice Numbers"

Practice
Use the Dot Set Cards to write an addition problem. Find the sum.

 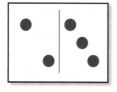 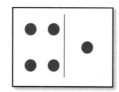   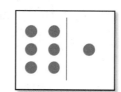 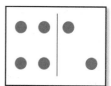

___ + ___ + ___ + ___ = ___   ___ + ___ + ___ + ___ = ___

Use two Dot Set Cards to create an addition sentence that has each sum below. Draw the Dot Set Cards you used to create the sum, and then write the addition sentence below.

9 12 =  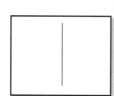   **10** 14 =

___ + ___ + ___ + ___ = 12   ___ + ___ + ___ + ___ = 14

Reflect
Use Dot Set Cards to write two number sentences that have a sum of 15. Create one using the doubles-facts strategy and one using the "nice-numbers" strategy.

1. Doubles Fact: 15 = ___ + ___ + ___ + ___
2. Regrouping with "Nice Numbers": 15 = ___ + ___ + ___ + ___

Using Grouping • Lesson 2

Week 3 — Using Grouping

Lesson 3

Key Idea

You can use more than one strategy when adding several numbers.

Doubles Facts
$4 + 7 + 4 =$
$(4 + 4) + 7 =$
$8 + 7 =$
15

Near-Doubles Facts
$9 + 4 + 5 =$
$9 + (5 + 5 - 1) =$
$9 + (10 - 1) =$
18

"Nice Numbers"
$6 + 3 + 4 =$
$(6 + 4) + 3 =$
$10 + 3 =$
13

Try This

Write the sums of both number sentences. Determine whether the calculator input is a correct way of finding it.

1 Original number sentence: $7 + 3 + 5 + 4 =$ _____

Calculator input: $10 + 9 =$ _____

Does it work? _____

2 Original number sentence: $4 + 3 + 10 + 4 =$ _____

Calculator input: $10 + 8 + 4 =$ _____

Does it work? _____

3 Original number sentence: $8 + 4 + 2 =$ _____

Calculator input: $10 + 6 =$ _____

Does it work? _____

4 Original number sentence: $9 + 4 + 7 + 3 =$ _____

Calculator input: $10 + 3 + 10 =$ _____

Does it work? _____

⑤ Original number sentence: 4 + 3 + 8 + 3 = _____

Calculator input: 8 + 8 + 2 = _____

Does it work? _____

⑥ Original number sentence: 3 + 8 + 3 + 2 = _____

Calculator input: 10 + 9 = _____

Does it work? _____

Practice
Find each sum. Write the strategy or strategies that you used.

⑦ 10 + 7 + 2 = _____

⑧ 7 + 7 + 2 + 5 = _____

⑨ 3 + 9 + 3 + 9 = _____

⑩ 4 + 6 + 3 + 6 = _____

Reflect
Write a number string of 4 numbers for each sum.

21 _____

18 _____

Using Grouping • Lesson 3

Week 3 — Using Grouping

Lesson 4

> **Key Idea**
> Watch out for numbers in the word problem that are not related to what the question is asking.

Try This
Solve each word problem using addition. Write the number sentence.

1. Kevyn went to a carnival three days in a row. On Friday he rode 4 rides. On Saturday he rode 6 rides. On Sunday he rode 4 rides. How many rides did Kevyn ride in all?

   _____

2. Harold wanted to win tickets from the sports booth. He bought 3 tickets the first round, but he did not win. He bought 5 tickets the second round, but he did not win. Then he bought 7 tickets the third round and finally won. How many total tickets did Harold buy?

   _____

3. Elizabeth sold 12 hot dogs during the first hour working at the food stand. During the second hour, she sold 11 hot dogs. During the third and fourth hours, she sold 8 and 10 hot dogs. How many total hot dogs did Elizabeth sell during her four-hour shift?

   _____

Practice
Solve each word problem using addition. Write the number sentence.

❹ A clown making balloon animals used 4 balloons to make a dog. He used 7 balloons to create a monkey. The giraffe he made required 5 balloons. He made a bird using 4 balloons. How many total balloons did the clown use to make these animals?

_____

❺ Margaret started selling tickets for the cake booth at 1:00. She sold the first set in 8 minutes. It took her 5 minutes to sell the second set and 9 minutes to sell the third set. How long did it take for Margaret to sell the three sets of tickets?

_____

Reflect
Angie solved the following problem.

Tyrese bought fish for his fish tank three days in a row this week. On Monday he bought 7 fish. On Tuesday he bought 8 fish. On Wednesday he bought 5 fish. How many fish did Tyrese buy in all?

Angie wrote 3 + 7 + 8 + 5 = 23 fish. Is she correct? If not, what should she have written?

_____

_____

_____

_____

Using Grouping

Lesson 5 Review

This week you explored more addition strategies. You learned that the order of the addends does not affect the sum, so you can reorder the addends to use "nice numbers," doubles facts, and near-doubles facts.

Lesson 1 Answer the following questions.

❶ Does 8 + 3 = 3 + 8? _____   ❷ Does 7 + 5 = 5 + 7? _____

If yes, what is the sum? _____   If yes, what is the sum? _____

❸ What do you notice about the two problems above?

_____

Lesson 2 Write two number strings for the sum. Use the doubles-facts strategy for one and the "nice-numbers" strategy for the other.

❹ 15

Doubles Fact: _____

"Nice Number": _____

Reflect
Show how to reorder the addends to use doubles facts in the problem below. Find the sum. Then show how to reorder the addends to use "nice numbers." Find the sum.

7 + 8 + 3 + 7

_____

_____

Lesson 3 Reorder the addends in the addition problem to make "nice numbers." Then find the sum.

⑤ 6 + 9 + 17 + 14 + 3

(___ + ___) + (___ + ___) + ___

___ + ___ + ___ = ___

Lesson 4 Solve each word problem using addition. Write the number sentence.

⑥ Brittany and Kimberly played **Race for the Sum**. Their first round took 12 minutes. The second round took 16 minutes, and the third round was over in 8 minutes. The final round was finished in 10 minutes. How long did the girls play?

_____

⑦ John was in charge of setting up the carnival booths. There were three rows of booths to set up. The first row took 20 minutes, and the second row took 18 minutes. John needed 22 minutes for the third row. How long did John take to put up all the booths?

_____

Reflect
Explain why addends were reordered in the problem below. Is the sum correct? If not, correct the mistake.

22 + 9 + 13 + 8 + 7   (22 + 8) + (13 + 7) + 9

20 + 20 + 9 = 49

_____
_____
_____

Using Grouping • Lesson 5 Review

Week 4 — Using Partial Sums

Lesson 1

> **Key Idea**
> Models and pictures can help with renaming when adding two-digit numbers.

Try This
Circle each set of 10 unit blocks below. Redraw the value given using rods and unit blocks. Write the value.

Practice

Use Base-Ten Blocks to create a model for each number. Rename ones to tens as needed. Find the sum.

⑤   62
   + 29

⑥   45
   + 37

⑦   39
   + 14

⑧   77
   + 18

Draw a picture using Base-Ten Blocks for each problem. Find the sum.

⑨   49
   + 47

⑩   67
   + 18

Reflect

Look at the following problem. Is the answer correct? Explain.

  59
+ 46
―――
 915

_____

_____

_____

Using Partial Sums • Lesson 1

Week 4 — Using Partial Sums

Lesson 2

Key Idea
Models and pictures can help you add three-digit numbers with renaming.

Try This
Circle each set of 10 unit blocks below. Circle each set of 10 rods below. Redraw the value given using flats, rods, and unit blocks. Write the value.

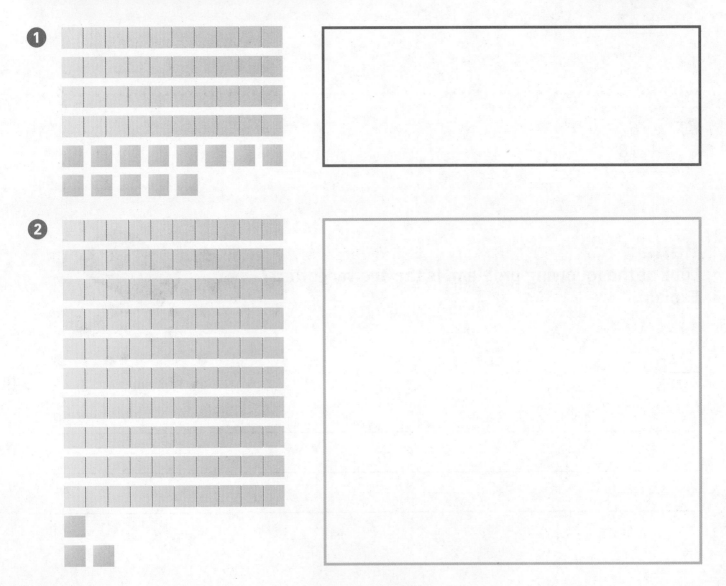

Practice
Use Base-Ten Blocks to create a model for each number. Rename ones to tens, and tens to hundreds as needed. Find the sum.

③ 137
  + 325

④ 275
  + 142

⑤ 435
  + 236

⑥ 378
  + 163

Find the sum.

⑦ 639
  + 145

⑧ 434
  + 567

Reflect
Explain the renaming that you did in Problem 8.

_____
_____
_____
_____

Week 4 — Using Partial Sums

Lesson 3

Key Idea

Find the sum of the tens digits and the sum of the ones digits, and then add them. This strategy is called *partial sums*.

```
   27
  +46
   60   (20 + 40) Add tens.
  +13   (7 + 6) Add ones.
   73
```

Try This

Write each number as the sum of tens and ones values.

1. 46 = ____ + ____

2. 71 = ____ + ____

3. 35 = ____ + ____

4. 62 = ____ + ____

Fill in the missing values using the partial-sums strategy.

5.
```
      35
  +   17
  ____   (30 + 10)
  +____  (5 + 7)
```

6.
```
      64
  +   21
  ____   (60 + 20)
  +____  (4 + 1)
```

7.
```
      32
  +   56
      80  (____ + ____)
  +    8  (____ + ____)
```

8.
```
      49
  +   15
      50  (____ + ____)
  +   14  (____ + ____)
```

36 Addition • Week 4

Practice
Fill in the missing values using the partial-sums strategy.

9)
```
   26
+  43
```
____ (____ + ____)
+____ (____ + ____)

10)
```
   13
+  39
```
____ (____ + ____)
+____ (____ + ____)

Find each sum using the partial-sums strategy.

11)
```
   70
+  22
```

12)
```
   54
+  21
```

13)
```
   28
+  63
```

14)
```
   36
+  43
```

Reflect
Use the partial-sums strategy to find the sum. Explain each of your steps.

```
   54
+  38
```

_____
_____
_____

Using Partial Sums • Lesson 3

Week 4 — Using Partial Sums

Lesson 4

Key Idea

The partial-sums strategy can also be used with three-digit numbers.

```
    314
  + 158
  -----
    400   (300 + 100) Add hundreds.
     60   (10 + 50)  Add tens.
  +  12   (4 + 8)    Add ones.
  -----
    472
```

Try This

Write each number as the sum of hundreds, tens, and ones values.

1 182 = _____ + _____ + _____

2 719 = _____ + _____ + _____

3 304 = _____ + _____

4 520 = _____ + _____

Fill in the missing values using the partial-sums strategy.

5
```
    243
  + 151
```
_____ (200 + 100)

_____ (40 + 50)

+_____ (3 + 1)

6
```
    126
  + 315
```
_____ (100 + 300)

_____ (20 + 10)

+_____ (6 + 5)

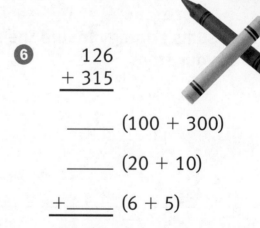

Practice

Fill in the missing values using the partial-sums strategy.

7. 436
    + 261

___ (___ + ___)
___ (___ + ___)
+ ___ (___ + ___)

8. 753
    + 138

___ (___ + ___)
___ (___ + ___)
+ ___ (___ + ___)

Find each sum using the partial-sums strategy.

9. 642
    + 153

10. 320
     + 246

11. 343
     + 111

12. 227
     + 516

Reflect

Use the partial-sums strategy to find the sum. Explain each of your steps.

  743
+ 352

_____

_____

_____

_____

Using Partial Sums • Lesson 4

Using Partial Sums

Lesson 5 Review

This week you explored more addition strategies. You used partial sums and Base-Ten Blocks with renaming to find sums.

Lesson 1   Use the partial-sums strategy to find each sum.

1
```
   38
 + 55
```
____ (30 + 50)

+____ (8 + 5)

2
```
   138
 + 312
```
____ (100 + 300)

____ (30 + 10)

+____ (8 + 2)

Lesson 2   Use the partial-sums strategy to find each sum.

3
```
   37
 + 44
```
____ (____ + ____)

+____ (____ + ____)

4
```
   536
 + 368
```
____ (____ + ____)

____ (____ + ____)

+____ (____ + ____)

Reflect

Name the partial sums for the problem below. Find the sum.

```
   556
 + 245
```

40   Addition • Week 4

Lesson 3 Use Base-Ten Blocks to create a model for each number. Rename ones to tens and tens to hundreds as needed. Draw your model. Find the sum.

⑤ 64
  + 27

⑥ 34
  + 19

⑦ 78
  + 17

Lesson 4 ## Practice
Use Base-Ten Blocks to create a model for each number. Rename ones to tens and tens to hundreds as needed. Find the sum.

⑧ 177
  + 286

⑨ 454
  + 277

⑩ 194
  + 268

Reflect
Which problem below does not require any renaming to find the sum? Explain.

⑪ 31
  + 55

⑫ 268
  + 513

⑬ 344
  + 292

_____

_____

Using Partial Sums • Lesson 5 Review 41

Week 1 — Using Doubles and Near Doubles

Practice

Use doubles facts to find the sums of each near-doubles fact.

1. If 20 + 20 = ____, then
   20 + 21 = ____.

2. If 55 + 55 = ____, then
   55 + 56 = ____.

Use doubles facts to find the missing addend.

3. 12 + ____ = 23

4. 14 + ____ = 29

Use the Number Construction Mat and Base-Ten Blocks to create a model of the addition problems. Find each sum.

5. 14 + 10 = ____

6. 46 + 10 = ____

Using a 100 Chart, find each sum.

7. 47 + 9 = ____

8. 19 + 9 = ____

Use +10 combinations to help you decide if the missing addend is 9 or 11. Circle your choice.

9. 18 + (9 or 11) = 29

10. 22 + (9 or 11) = 31

Using doubles facts, near-doubles facts, +10 combinations, +9 combinations, or +11 combinations, make an addition problem for the following sum. Write the strategy used.

11. 39 = ____ + ____   _____

42 Addition • Week 1 Practice

Week 2 — Using Nice Numbers

Practice

Circle the numbers below that are considered "nice numbers."

① 18   40   96   11   29   70   35   10   24   33   64

Write a new addition sentence by sharing the values. Be sure that a "nice number" is in the new sentence.

② 56 + 41

New addition sentence

_____

③ 17 + 62

New addition sentence

_____

Use a Number Construction Mat and Base-Ten Blocks or a Place Value Mat to build each addition problem. Say how many tens and ones are modeled.

④ 23 + 74

____ tens and ____ ones = ____

⑤ 46 + 12

____ tens and ____ ones = ____

⑥ 51 + 18

____ tens and ____ ones = ____

⑦ 36 + 41

____ tens and ____ ones = ____

Regroup each number into tens and ones. Then find each sum.

⑧   42 = 40 + 2
     + 16 = 10 + 6
     ─────────────
     _____

⑨   26 = _____
     + 61 = _____
     ─────────────
     _____

Addition • Week 2 Practice   43

Week 3 Using Grouping

Practice

Answer the following questions.

1. Does 6 + 4 = 4 + 6? _____   If yes, what is the sum? _____

2. What do you notice about the problem above?
   _____

Reorder the numbers in each addition problem to make "nice numbers." Then find the sum.

3. 3 + 6 + 7 = _____

4. 9 + 8 + 2 + 1 = _____

Reorder the numbers in the addition problem to make "nice numbers." Then find the sum.

5. 7 + 5 + 13 + 4 + 5

   (____ + ____) + (____ + ____) + ____

   ____ + ____ + ____ = ____

Solve the word problem using addition. Write the number sentence.

6. Travis sets aside time for reading every night. Four nights ago, Travis read 26 pages of a book. He read 31 pages three nights ago and 21 pages two nights ago. Finally he read 30 pages last night. How many pages did Travis read over the last four nights?

   _____

44 Addition • Week 3 Practice

Week 4 — Using Partial Sums

Practice

Use the partial-sums strategy to find each sum.

1.
284
+ 685
_____ (200 + 600)
_____ (80 + 80)
+_____ (4 + 5)

2.
158
+ 421
500 (____ + ____)
70 (____ + ____)
+ 9 (____ + ____)

Use the partial-sums strategy to find each sum.

3.
46
+ 13
_____ (____ + ____)
+_____ (____ + ____)

4.
824
+ 115
_____ (____ + ____)
_____ (____ + ____)
+_____ (____ + ____)

Use Base-Ten Blocks to create a model for each number. Rename ones to tens and tens to hundreds as needed. Draw your model. Find the sum.

5. 26
+ 37

6. 14
+ 38

7. 629
+ 187

8. 557
+ 385

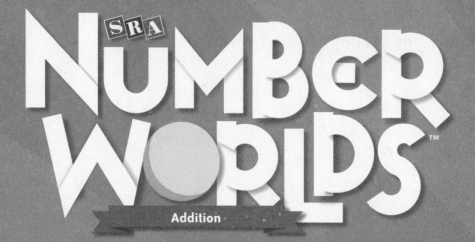

Addition

Unit 3 Workbook

SRAonline.com

Level E

Author
Sharon Griffin
*Associate Professor of Education and
Adjunct Associate Professor of Psychology*
Clark University
Worcester, Massachusetts

Building Blocks Authors

Douglas H. Clements
*Professor of Early Childhood
and Mathematics Education*
University at Buffalo
State University of New York, New York

Julie Sarama
Associate Professor of Mathematics Education
University at Buffalo
State University of New York, New York

Contributing Writers
Sherry Booth, *Math Curriculum Developer,* Raleigh, North Carolina
Elizabeth Jimenez, *English Language Learner Consultant,* Pomona, California

Program Reviewers

Jean Delwiche
Almaden Country School
San Jose, California

Cheryl Glorioso
Santa Ana Unified School District
Santa Ana, California

Sharon LaPoint
School District of Indian River County
Vero Beach, Florida

Leigh Lidrbauch
Pasadena Independent School District
Pasadena, Texas

Dave Maresh
Morongo Unified School District
Yucca Valley, California

Mary Mayberry
Mon Valley Education Consortium, AIU 3
Clairton, Pennsylvania

Lauren Parente
Mountain Lakes School District
Mountain Lakes, New Jersey

Juan Regalado
Houston Independent School District
Houston, Texas

M. Kate Thiry
Dublin City School District
Dublin, Ohio

Susan C. Vohrer
Baltimore County Public Schools
Baltimore, Maryland

SRAonline.com

Copyright © 2007 SRA/McGraw-Hill.

All rights reserved. Except as permitted under the United States Copyright Act, no part of this publication may be reproduced or distributed in any form or by any means, or stored in a database or retrieval system, without the prior written permission of the publisher, unless otherwise indicated.

Printed in the United States of America.

Send all inquiries to:
SRA/McGraw-Hill
4400 Easton Commons
Columbus, OH 43219-6188

R5313X.01

9 WCE 12 11 10

Photo Credits

3-39 ©PhotoDisc/Getty Images, Inc.

Contents

Addition

Week 1　Using Doubles and Near Doubles **2**

Week 2　Using Nice Numbers **12**

Week 3　Using Grouping **22**

Week 4　Using Partial Sums **32**

Week 1　Practice **42**

Week 2　Practice **43**

Week 3　Practice **44**

Week 4　Practice **45**

Week 1 — Using Doubles and Near Doubles

Lesson 1

Key Idea

2 + 2 = 4   Doubles Fact
2 + 3 = 2 + (2 + 1)   Near-Doubles Fact
     = 4 + 1 = 5

Try This

Find the sum of each doubles fact.

① 2 + 2 = ____

② 5 + 5 = ____

③ 4 + 4 = ____

④ 1 + 1 = ____

Use a doubles fact to find the sum of each near-doubles fact.

⑤ 2 + 3 = 2 + (2 + 1)
____ + 1 = ____

⑥ 5 + 6 = 5 + (5 + 1)
____ + 1 = ____

⑦ 3 + 4 = 3 + (3 + 1)
____ + 1 = ____

⑧ 4 + 5 = 4 + (4 + 1)
____ + 1 = ____

Practice

Use the sum of each doubles fact to find the sum of each near-doubles fact.

9 If 3 + 3 = _____, then

3 + 4 = _____.

10 If 5 + 5 = _____, then

5 + 6 = _____.

11 If 7 + 7 = _____, then

7 + 8 = _____.

12 If 8 + 8 = _____, then

8 + 9 = _____.

13 If 5 + 5 = _____, then

5 + 4 = _____.

14 If 2 + 2 = _____, then

1 + 2 = _____.

15 If 7 + 7 = _____, then

6 + 7 = _____.

16 If 4 + 4 = _____, then

3 + 4 = _____.

Find each sum.

17 3 + 4 = _____

18 5 + 6 = _____

19 10 + 9 = _____

20 7 + 6 = _____

21 8 + 7 = _____

22 5 + 4 = _____

Reflect

Given the problem 4 + 5, show how each of the doubles facts below can be used to find the sum.

4 + 4   or   5 + 5

_____ or _____

Using Doubles and Near Doubles • Lesson 1

Using Doubles and Near Doubles

Lesson 2

Key Idea

When 10 is added to a number, the digit in the tens place increases by 1. You can use a 99 Chart to add by circling the starting number, using an arrow to show a jump of 10, and shading in the sum.

12 + 10 = 22

21	22	23	24	25	26	27	28	29	30
11	(12)	13	14	15	16	17	18	19	20

Try This

Use the Number Construction Mat and Base-Ten Blocks to create a model of the addition problems. Find each sum.

1. 20 + 10 = _____

2. 50 + 10 = _____

3. 24 + 10 = _____

4. 38 + 10 = _____

Use a 99 Chart to find each sum.

5. 80 + 10 = _____

6. 60 + 10 = _____

7. 51 + 10 = _____

8. 72 + 10 = _____

9. In what direction do you move on a 99 Chart when 10 is added to a number?

_____

Practice
Use the Number Construction Mat and Base-Ten Blocks to create a model of the addition problems. Find each sum.

⑩ 13 + 10 = _____

⑪ 49 + 10 = _____

⑫ 36 + 10 = _____

⑬ 65 + 10 = _____

Use a 99 Chart to find each missing addend.

⑭ _____ + 10 = 40

⑮ _____ + 10 = 61

⑯ _____ + 10 = 66

⑰ _____ + 10 = 94

Use the Number Construction Mat and Base-Ten Blocks to create a model of the addition problems. Find each missing addend.

⑱ _____ + 10 = 29

⑲ _____ + 10 = 77

⑳ _____ + 10 = 37

㉑ _____ + 10 = 55

Reflect
What pattern do you notice when 10 is added to any number?
_____

Using Doubles and Near Doubles • Lesson 2

Week 1 — Using Doubles and Near Doubles

Lesson 3

Key Idea

On a 99 Chart:
Add 10 .... Move UP ↑
Add 1 .... Move RIGHT →
Subtract 1 ...... Move LEFT ←

Try This

Use a 99 Chart to find each sum or difference. Circle the correct direction arrow on the 99 Chart to find the sum or difference.

① 51 + 10 = _____   ↑ → ←

② 51 + 1 = _____   ↑ → ←

③ 51 − 1 = _____   ↑ → ←

④ 47 + 10 = _____   ↑ → ←

⑤ 47 + 1 = _____   ↑ → ←

⑥ 47 − 1 = _____   ↑ → ←

Use a 99 Chart to model each problem. Find each sum. Circle the correct direction arrow or arrows on a 99 Chart to find the sum.

⑦ 58 + 10 = _____   ↑ → ←

⑧ 63 + 11 =
  63 + (10 + 1) = _____   ↑ → ←

⑨ 15 + 9 =
  15 + (10 − 1) = _____   ↑ → ←

6 Addition • Week 1

Practice

Use a 99 Chart to model each problem. Find each sum and circle the direction or directions that you moved on the 99 Chart.

⑩ 16 + 10 = _____
   Start at 16 and move **up or down** 1 block.

⑪ 16 + 9 = _____
   Start at 16 and move **up or down** 1 block and **left or right** 1 block.

⑫ 16 + 11 = _____
   Start at 16 and move **up or down** 1 block and **left or right** 1 block.

⑬ 34 + 10 = _____
   Start at 34 and move **up or down** 1 block.

⑭ 34 + 9 = _____
   Start at 34 and move **up or down** 1 block and **left or right** 1 block.

⑮ 34 + 11 = _____
   Start at 34 and move **up or down** 1 block and **left or right** 1 block.

Reflect

What do you do differently using a 99 Chart when adding 9 rather than 11 to a number?

_____

_____

_____

Week 1 — Using Doubles and Near Doubles

Lesson 4

Key Idea

$10 + 10 = 20$

$10 + 9 = 19$
$10 + (10 - 1) = 19$
$20 - 1 = 19$

$10 + 11 = 21$
$10 + (10 + 1) = 21$
$20 + 1 = 21$

Try This
Find the sum of each doubles fact.

① $3 + 3 = $ _____

② $30 + 30 = $ _____

③ $4 + 4 = $ _____

④ $40 + 40 = $ _____

Use a doubles fact to find the sum of each near-doubles fact.

⑤ $8 + 9 = 8 + (8 + 1)$
_____ $+ 1 = $ _____

⑥ $8 + 7 = 8 + (8 - 1)$
_____ $- 1 = $ _____

⑦ $20 + 21 = 20 + (20 + 1)$
_____ $+ 1 = $ _____

⑧ $20 + 19 = 20 + (20 - 1)$
_____ $- 1 = $ _____

⑨ $40 + 41 = 40 + (40 + 1)$
_____ $+ 1 = $ _____

⑩ $40 + 39 = 40 + (40 - 1)$
_____ $- 1 = $ _____

⑪ $25 + 26 = 25 + (25 + 1)$
_____ $+ 1 = $ _____

⑫ $25 + 24 = 25 + (25 - 1)$
_____ $- 1 = $ _____

Practice

Use the sum of each doubles fact to find the sum of each near-doubles fact.

⑬ If 30 + 30 = _____, then

30 + 31 = _____.

⑭ If 15 + 15 = _____, then

15 + 14 = _____.

⑮ If 30 + 30 = _____, then

30 + 29 = _____.

⑯ If 15 + 15 = _____, then

15 + 16 = _____.

⑰ If 50 + 50 = _____, then

50 + 49 = _____.

⑱ If 45 + 45 = _____, then

45 + 46 = _____.

Use doubles facts to find each missing addend.

⑲ 7 + _____ = 15

⑳ 6 + _____ = 13

㉑ 30 + _____ = 61

㉒ 25 + _____ = 49

Reflect

Explain how using a doubles fact helps you find the sum of a near-doubles fact.

_____

_____

_____

Using Doubles and Near Doubles • Lesson 4

Week 1

Using Doubles and Near Doubles

Lesson 5 Review

This week you explored addition strategies. You discovered that doubles facts are helpful tools when working with near-doubles facts.

Lesson 1 Use a doubles fact to find the sum of each near-doubles fact.

① If 40 + 40 = _____, then 40 + 41 = _____.

② If 25 + 25 = _____, then 25 + 24 = _____.

③ If 35 + 35 = _____, then 35 + 36 = _____.

④ If 50 + 50 = _____, then 49 + 50 = _____.

Lesson 2 Use near-doubles facts to find each missing addend.

⑤ 7 + _____ = 13

⑥ 9 + _____ = 17

⑦ 30 + _____ = 61

⑧ 15 + _____ = 29

Use the Number Construction Mat and Base-Ten Blocks to create a model of the addition problems. Find each sum.

⑨ 16 + 10 = _____

⑩ 10 + 52 = _____

⑪ 43 + 10 = _____

⑫ 88 + 10 = _____

Reflect

Explain how to use a 99 Chart to find the sum of 38 + 9.

_____

_____

Lesson 3 Using a 99 Chart, find each sum.

⑬ 54 + 9 = _____   ⑭ 87 + 9 = _____

⑮ 38 + 11 = _____   ⑯ 19 + 11 = _____

⑰ 22 + 9 = _____   ⑱ 61 + 9 = _____

Use +10 combinations to help you decide if the missing addend is 9 or 11. Circle your choice.

⑲ 15 + (9 or 11) = 26   ⑳ 55 + (9 or 11) = 66

㉑ 34 + (9 or 11) = 43   ㉒ 46 + (9 or 11) = 57

㉓ 79 + (9 or 11) = 88   ㉔ 58 + (9 or 11) = 67

Lesson 4 Using doubles facts, near-doubles facts, +10 combinations, +9 combinations, or +11 combinations, make an addition problem for the following sums. Use a different strategy for each. Write the strategy used on the line provided.

㉕ 54 = _____ + _____   _____

㉖ 48 = _____ + _____   _____

㉗ 66 = _____ + _____   _____

㉘ 39 = _____ + _____   _____

Reflect
Which strategies do you like to use best? Why?

_____

_____

_____

Week 2

Using Nice Numbers

Lesson 1

Key Idea

6 + 4 = 10

18 + 2 = 20

A "nice number" is a number that ends in a 0.

Try This

On the number line, circle all the "nice numbers."

1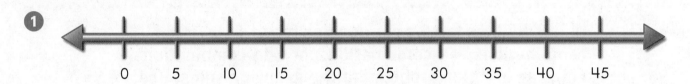

Using the number line above, find the given number's closest "nice number."

2 23 _____   **3** 18 _____   **4** 41 _____

5 28 _____   **6** 37 _____   **7** 32 _____

Practice

Make a "nice number" by drawing the missing boxes. Find the missing addend.

8 7 + ____ = 10

9 ____ + 19 = 20

12 Addition • Week 2

Practice

Make a "nice number" by writing the missing addend.

10

_____ + 16 = 20

11

_____ + 5 = 10

Use the Number Construction Mat and Base-Ten Blocks to show a model of the following problems. Write an addition sentence using the "nice number" that is described.

12 19 + 3   Share 1 from 3 and add it to 19.

New addition sentence: _____

13 28 + 6   Share 2 from 6 and add it to 28.

New addition sentence: _____

Write a new addition sentence by sharing the values. Make sure that a "nice number" is in the new sentence.

14 13 + 6

New addition sentence: _____

15 34 + 11

New addition sentence: _____

Reflect

Give three addition problems that have a sum of 24. Then underline the number in each sentence that is closer to a nice number.

Week 2 — **Using Nice Numbers**

Lesson 2

Key Idea

You can regroup two-digit numbers into "nice numbers" to make them easier to work with.

24 = 20 + 4

Tens	Ones
(2 rods)	(4 units)

↑ Its value is 20. ↑ Its value is 4.

Try This

Use a Number Construction Mat and Base-Ten Blocks or a Place Value Mat to build each number. Say how many tens and ones are shown.

① 24 ____ tens and ____ ones

② 12 ____ tens and ____ ones

③ 35 ____ tens and ____ ones

④ 29 ____ tens and ____ ones

Use a Number Construction Mat and Base-Ten Blocks or a Place Value Mat to build each addition problem. Say how many tens and ones are shown.

⑤ 21 + 11

____ tens and ____ ones = ____

⑥ 24 + 15

____ tens and ____ ones = ____

⑦ 16 + 13

____ tens and ____ ones = ____

⑧ 17 + 12

____ tens and ____ ones = ____

Practice

Regroup each number into tens and ones.

⑨ 16 = ____ + ____

⑩ 13 = ____ + ____

⑪ 11 = ____ + ____

⑫ 12 = ____ + ____

Regroup each number into tens and ones.
Then find each sum.

13. 16 = 10 + 6
     + 13 = 10 + 3
     _____
            20 + 9 = ____

14. 22 = 20 + 2
     + 17 = 10 + 7
     _____
            ____ + ____ = ____

15. 17 = ____ + ____
     + 12 = ____ + ____
     _____
            ____ + ____ = ____

16. 11 = ____ + ____
     + 24 = ____ + ____
     _____
            ____ + ____ = ____

17. 13 = ____ + ____
     + 25 = ____ + ____
     _____
            ____ + ____ = ____

18. 19 = ____ + ____
     + 16 = ____ + ____
     _____
            ____ + ____ = ____

19. 17 = ____ + ____
     + 11 = ____ + ____
     _____
            ____ + ____ = ____

20. 27 = ____ + ____
     + 13 = ____ + ____
     _____
            ____ + ____ = ____

Reflect
Using a Number Construction Mat and Base-Ten Blocks, draw a model of an addition problem that has a sum of 47.

Tens	Ones

Using Nice Numbers • Lesson 2

Week 2 — Lesson 3

Using Nice Numbers

Key Idea

Regrouping numbers into tens and ones is a strategy for adding two-digit numbers.

$$23 = 20 + 3$$
$$+ 46 = 40 + 6$$
$$60 + 9 = 69$$

Try This

Use a Number Construction Mat and Base-Ten Blocks or a Place Value Mat to build the following numbers. Show how many tens and ones you made.

1. 62 _____ tens and _____ ones
2. 74 _____ tens and _____ ones
3. 48 _____ tens and _____ ones
4. 95 _____ tens and _____ ones

Rewrite each number as tens and ones.

5. 56 = _____ + _____
6. 92 = _____ + _____
7. 64 = _____ + _____
8. 18 = _____ + _____

Practice

Use the Place Value Mat to build each addition problem. Show how many tens and ones are in the sum.

9. 28 + 31

    _____ tens and _____ ones = _____

10. 52 + 15

    _____ tens and _____ ones = _____

11. 77 + 12

    _____ tens and _____ ones = _____

12. 64 + 22

    _____ tens and _____ ones = _____

16 Addition • Week 2

Regroup each number into tens and ones.
Then find each sum.

13 54 = ____ + ____
   + 23 = ____ + ____
   _____
       ____ + ____ = ____

14 18 = ____ + ____
   + 71 = ____ + ____
   _____
       ____ + ____ = ____

15 66 = ____ + ____
   + 32 = ____ + ____
   _____
       ____ + ____ = ____

16 84 = ____ + ____
   + 13 = ____ + ____
   _____
       ____ + ____ = ____

Regroup to find each sum.

17   62
     + 27
     ____

18   43
     + 25
     ____

19   51
     + 18
     ____

20   23
     + 36
     ____

21   48
     + 10
     ____

22   74
     + 26
     ____

Reflect
Jackie thinks it is easier to add 50 + 40 + 6 + 3 than 56 + 43. Do you agree with her? Why or why not?

_____

_____

Using Nice Numbers • Lesson 3

Week 2 — Using Nice Numbers
Lesson 4

Key Idea
"Leaps of 10" is an addition strategy that uses a number line to find the sum.

Use the following steps to help you find sums.
Step 1 Locate the first number on a number line.
Step 2 Break the second number into groups of tens and ones.
Step 3 Make "leaps of 10" and "hops of 1."

Try This
Write how many leaps and hops you make for each number.

1. 34 Leaps _____ Hops _____
2. 7   Leaps _____ Hops _____
3. 49 Leaps _____ Hops _____
4. 25 Leaps _____ Hops _____
5. 36 Leaps _____ Hops _____
6. 12 Leaps _____ Hops _____

Practice
Break down the following numbers into tens and ones.

7. 52 = _____ tens, _____ ones = 10 + 10 + 10 + 10 + 10 + 1 + 1
8. 44 = _____ tens, _____ ones = _____
9. 39 = _____ tens, _____ ones = _____
10. 21 = _____ tens, _____ ones = _____

Make leaps and hops on the number line to find each sum. Complete the addition sentences.

⑪ 20 + 10 = _____

⑫ 47 + 10 = _____

⑬ 54 + 20 = 54 + _____ + _____ = _____

⑭ 16 + 41 = 16 + _____ + _____ + _____ + _____ + _____ = _____

⑮ 62 + 34 = _____ = _____

⑯ 29 + 22 = _____ = _____

Reflect
Monica says that when you add 42 + 23, you need to make two moves on the number line to find the sum. Is she correct? Explain.

_____

_____

_____

_____

What is the sum of 42 + 23? _____

Using Nice Numbers • Lesson 4

Week 2 — Using Nice Numbers

Lesson 5 Review

This week you explored addition strategies. You used "nice numbers," regrouping, and "leaps of 10" to find the sums.

Lesson 1 Circle the numbers below that are considered "nice numbers."

① 20   31   55   7   94   60   73   68   3   10   19

Write a new addition sentence by sharing the values. Make sure the new sentence includes a "nice number."

② 17 + 8

New addition sentence:
_____

③ 22 + 17

New addition sentence:
_____

Lesson 2 Use a Number Construction Mat and Base-Ten Blocks or a Place Value Mat to build each addition problem. Write how many tens and ones are shown.

④ 41 + 33 ____ tens and ____ ones = ____

⑤ 28 + 31 ____ tens and ____ ones = ____

⑥ 17 + 22 ____ tens and ____ ones = ____

⑦ 63 + 25 ____ tens and ____ ones = ____

Reflect
Explain the sharing you did to write the new addition sentence in Problem 3.

_____

_____

Lesson 3   Regroup each number into tens and ones. Then find each sum.

8.   43 = 40 + 3
+ 32 = 30 + 2
_____

_____

9.   15 = _____
+ 63 = _____

_____

10.   13 = _____
+ 25 = _____

_____

11.   51 = _____
+ 46 = _____

_____

Lesson 4   Make leaps and hops on the number line to find each sum.

12.   23 + 43 = 23 + 10 + 10 + 10 + 10 + 1 + 1 + 1

= ____

13.   51 + 34 = 51 + 10 + 10 + 10 + 1 + 1 + 1 + 1 = ____

Reflect

Draw a number line to illustrate the following addition problem using "leaps of 10" and "hops of 1." Explain your moves.

56 + 23

Week 3 — Using Grouping

Lesson 1

Key Idea
When adding, the order of the numbers does not matter.

1 2 3 4 5 6 7 8 9 10

6 + 4 = 10

1 2 3 4 5 6 7 8 9 10

4 + 6 = 10

Try This
Find each sum.

① 8 + 3 = _____   3 + 8 = _____

② 5 + 7 = _____   7 + 5 = _____

③ 5 + 4 = _____   4 + 5 = _____

④ 20 + 10 = _____   10 + 20 = _____

Reorder the numbers in each addition problem to make "nice numbers." Then find the sum.

⑤ 6 + 3 + 4 = 6 + 4 + 3

_____ + _____ = _____

⑥ 9 + 5 + 1 = 9 + 1 + 5

_____ + _____ = _____

⑦ 13 + 7 + 4 = _____ + _____ + 4

_____ + _____ = _____

22 Addition • Week 3

Reorder and group the addends to make "nice numbers" to help you find each sum.

⑧ 15 + 3 + 8 + 7 + 5

(15 + ____) + (7 + ____) + 8

____ + ____ + ____ = ____

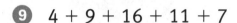

⑨ 4 + 9 + 16 + 11 + 7

(16 + ____) + (11 + ____) + ____

____ + ____ + ____ = ____

⑩ 21 + 15 + 5 + 7 + 9

(____ + ____) + (____ + ____) + ____

____ + ____ + ____ = ____

⑪ 8 + 13 + 7 + 12 + 5

(____ + ____) + (____ + ____) + ____

____ + ____ + ____ = ____

Reflect
Reorder and group the addends to make the problem easier to solve. Explain why this makes it easier to solve.

18 + 4 + 16 + 10 + 2

_____

_____

_____

Week 3 — Using Grouping

Lesson 2

Key Idea

Two strategies to make adding easier are the following:

1. **Reorder Using Doubles Facts:**
   $4 + 3 + 4 = (4 + 4) + 3 = 8 + 3 = 11$

2. **Reorder Using "Nice Numbers":**
   $2 + 5 + 8 = (8 + 2) + 5 = 10 + 5 = 15$

Try This

Write an addition sentence for each Dot Set Card.

1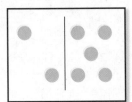

2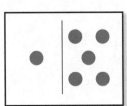

_____   _____

3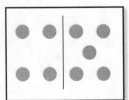

4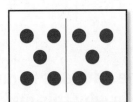

_____   _____

Use the Dot Set Cards to write an addition problem.
Find the sum. Circle the strategy you used.

5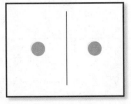

____ + ____ + ____ + ____ = ____

Doubles Fact   or   "Nice Numbers"

24 Addition • Week 3

6 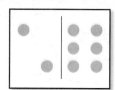 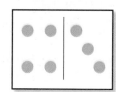   ___ + ___ + ___ + ___ = ___

Doubles Fact   or   "Nice Numbers"

Practice
Use the Dot Set Cards to write an addition problem. Find the sum.

7 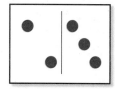 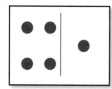   **8**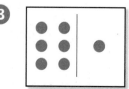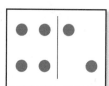

___ + ___ + ___ + ___ = ___   ___ + ___ + ___ + ___ = ___

Use two Dot Set Cards to create an addition sentence that has each sum below. Draw the Dot Set Cards you used to create the sum, and then write the addition sentence below.

9 12 =     **10** 14 = 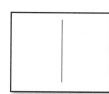

___ + ___ + ___ + ___ = 12   ___ + ___ + ___ + ___ = 14

Reflect
Use Dot Set Cards to write two number sentences that have a sum of 15. Create one using the doubles-facts strategy and one using the "nice-numbers" strategy.

1. Doubles Fact: 15 = ___ + ___ + ___ + ___
2. Regrouping with "Nice Numbers":   15 = ___ + ___ + ___ + ___

Using Grouping • Lesson 2

Week 3 — Using Grouping

Lesson 3

Key Idea

You can use more than one strategy when adding several numbers.

Doubles Facts	Near-Doubles Facts	"Nice Numbers"
$4 + 7 + 4 =$	$9 + 4 + 5 =$	$6 + 3 + 4 =$
$(4 + 4) + 7 =$	$9 + (5 + 5 - 1) =$	$(6 + 4) + 3 =$
$8 + 7 =$	$9 + (10 - 1) =$	$10 + 3 =$
15	18	13

Try This

Write the sums of both number sentences. Determine whether the calculator input is a correct way of finding it.

1 Original number sentence: $7 + 3 + 5 + 4 =$ _____

Calculator input: $10 + 9 =$ _____

Does it work? _____

2 Original number sentence: $4 + 3 + 10 + 4 =$ _____

Calculator input: $10 + 8 + 4 =$ _____

Does it work? _____

3 Original number sentence: $8 + 4 + 2 =$ _____

Calculator input: $10 + 6 =$ _____

Does it work? _____

4 Original number sentence: $9 + 4 + 7 + 3 =$ _____

Calculator input: $10 + 3 + 10 =$ _____

Does it work? _____

⑤ Original number sentence: $4 + 3 + 8 + 3 = $ _____

Calculator input: $8 + 8 + 2 = $ _____

Does it work? _____

⑥ Original number sentence: $3 + 8 + 3 + 2 = $ _____

Calculator input: $10 + 9 = $ _____

Does it work? _____

Practice
Find each sum. Write the strategy or strategies that you used.

⑦ $10 + 7 + 2 = $ _____

⑧ $7 + 7 + 2 + 5 = $ _____

⑨ $3 + 9 + 3 + 9 = $ _____

⑩ $4 + 6 + 3 + 6 = $ _____

Reflect
Write a number string of 4 numbers for each sum.

21 _____

18 _____

Using Grouping • Lesson 3

Week 3 — Lesson 4

Using Grouping

Key Idea
Watch out for numbers in the word problem that are not related to what the question is asking.

Try This
Solve each word problem using addition. Write the number sentence.

① Kevyn went to a carnival three days in a row. On Friday he rode 4 rides. On Saturday he rode 6 rides. On Sunday he rode 4 rides. How many rides did Kevyn ride in all?

_____

② Harold wanted to win tickets from the sports booth. He bought 3 tickets the first round, but he did not win. He bought 5 tickets the second round, but he did not win. Then he bought 7 tickets the third round and finally won. How many total tickets did Harold buy?

_____

③ Elizabeth sold 12 hot dogs during the first hour working at the food stand. During the second hour, she sold 11 hot dogs. During the third and fourth hours, she sold 8 and 10 hot dogs. How many total hot dogs did Elizabeth sell during her four-hour shift?

_____

Practice
Solve each word problem using addition. Write the number sentence.

④ A clown making balloon animals used 4 balloons to make a dog. He used 7 balloons to create a monkey. The giraffe he made required 5 balloons. He made a bird using 4 balloons. How many total balloons did the clown use to make these animals?

_____

⑤ Margaret started selling tickets for the cake booth at 1:00. She sold the first set in 8 minutes. It took her 5 minutes to sell the second set and 9 minutes to sell the third set. How long did it take for Margaret to sell the three sets of tickets?

_____

Reflect
Angie solved the following problem.

Tyrese bought fish for his fish tank three days in a row this week. On Monday he bought 7 fish. On Tuesday he bought 8 fish. On Wednesday he bought 5 fish. How many fish did Tyrese buy in all?

Angie wrote $3 + 7 + 8 + 5 = 23$ fish. Is she correct? If not, what should she have written?

_____

_____

_____

_____

Week 3 · Using Grouping

Lesson 5 Review

This week you explored more addition strategies. You learned that the order of the addends does not affect the sum, so you can reorder the addends to use "nice numbers," doubles facts, and near-doubles facts.

Lesson 1 Answer the following questions.

❶ Does 8 + 3 = 3 + 8? _____   ❷ Does 7 + 5 = 5 + 7? _____

If yes, what is the sum? _____   If yes, what is the sum? _____

❸ What do you notice about the two problems above?

_____

Lesson 2 Write two number strings for the sum. Use the doubles-facts strategy for one and the "nice-numbers" strategy for the other.

❹ 15

Doubles Fact: _____

"Nice Number": _____

Reflect

Show how to reorder the addends to use doubles facts in the problem below. Find the sum. Then show how to reorder the addends to use "nice numbers." Find the sum.

7 + 8 + 3 + 7

_____

_____

Lesson 3 Reorder the addends in the addition problem to make "nice numbers." Then find the sum.

⑤ 6 + 9 + 17 + 14 + 3

(___ + ___) + (___ + ___) + ___

___ + ___ + ___ = ___

· ·

Lesson 4 Solve each word problem using addition. Write the number sentence.

⑥ Brittany and Kimberly played **Race for the Sum.** Their first round took 12 minutes. The second round took 16 minutes, and the third round was over in 8 minutes. The final round was finished in 10 minutes. How long did the girls play?

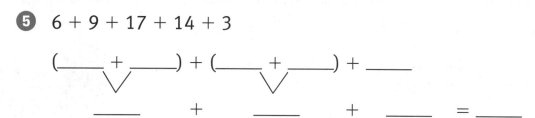

_____

⑦ John was in charge of setting up the carnival booths. There were three rows of booths to set up. The first row took 20 minutes, and the second row took 18 minutes. John needed 22 minutes for the third row. How long did John take to put up all the booths?

_____

Reflect
Explain why addends were reordered in the problem below. Is the sum correct? If not, correct the mistake.

22 + 9 + 13 + 8 + 7   (22 + 8) + (13 + 7) + 9

                              20   +   20   + 9 = 49

_____

_____

_____

Using Grouping • Lesson 5 Review

Using Partial Sums

Lesson 1

Key Idea
Models and pictures can help with renaming when adding two-digit numbers.

Try This
Circle each set of 10 unit blocks below. Redraw the value given using rods and unit blocks. Write the value.

32   Addition • Week 4

Practice

Use Base-Ten Blocks to create a model for each number. Rename ones to tens as needed. Find the sum.

5) 62
   + 29

6) 45
   + 37

7) 39
   + 14

8) 77
   + 18

Draw a picture using Base-Ten Blocks for each problem. Find the sum.

9) 49
   + 47

10) 67
    + 18

Reflect

Look at the following problem. Is the answer correct? Explain.

```
   59
 + 46
  915
```

_____

_____

_____

Using Partial Sums • Lesson 1

Using Partial Sums
Week 4 Lesson 2

> **Key Idea**
> Models and pictures can help you add three-digit numbers with renaming.

Try This

Circle each set of 10 unit blocks below. Circle each set of 10 rods below. Redraw the value given using flats, rods, and unit blocks. Write the value.

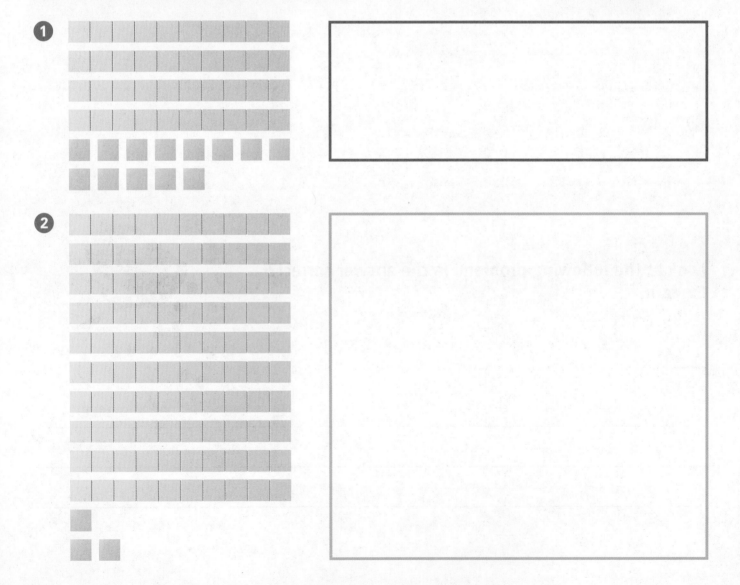

34 Addition • Week 4

Practice

Use Base-Ten Blocks to create a model for each number. Rename ones to tens, and tens to hundreds as needed. Find the sum.

3)   137
     + 325

4)   275
     + 142

5)   435
     + 236

6)   378
     + 163

Find the sum.

7)   639
     + 145

8)   434
     + 567

Reflect

Explain the renaming that you did in Problem 8.

_____

_____

_____

_____

Using Partial Sums • Lesson 2

Week 4 — **Using Partial Sums**

Lesson 3

Key Idea

Find the sum of the tens digits and the sum of the ones digits, and then add them. This strategy is called *partial sums*.

```
   27
  +46
   60   (20 + 40) Add tens.
  +13   (7 + 6) Add ones.
   73
```

Try This

Write each number as the sum of tens and ones values.

1. 46 = _____ + _____
2. 71 = _____ + _____
3. 35 = _____ + _____
4. 62 = _____ + _____

Fill in the missing values using the partial-sums strategy.

5.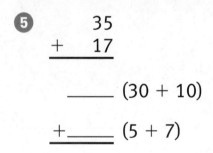
    ```
        35
     + 17
     _____ (30 + 10)
    +_____ (5 + 7)
    ```

6.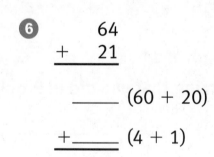
    ```
        64
     + 21
     _____ (60 + 20)
    +_____ (4 + 1)
    ```

7.  ```
         32
      +  56
         80 (____ + ____)
      +   8 (____ + ____)
    ```

8.  ```
         49
      +  15
         50 (____ + ____)
      +  14 (____ + ____)
    ```

36 Addition • Week 4

Practice
Fill in the missing values using the partial-sums strategy.

9. 26
     + 43
     ___ (___ + ___)
    +___ (___ + ___)

10. 13
      + 39
      ___ (___ + ___)
     +___ (___ + ___)

Find each sum using the partial-sums strategy.

11. 70
     + 22

12. 54
     + 21

13. 28
     + 63

14. 36
     + 43

Reflect
Use the partial-sums strategy to find the sum. Explain each of your steps.

  54
+ 38

Using Partial Sums • Lesson 3

Using Partial Sums

Lesson 4

Key Idea

The partial-sums strategy can also be used with three-digit numbers.

```
   314
 + 158
 ─────
   400   (300 + 100) Add hundreds.
    60   (10 + 50) Add tens.
 +  12   (4 + 8) Add ones.
 ─────
   472
```

Try This

Write each number as the sum of hundreds, tens, and ones values.

① 182 = _____ + _____ + _____

② 719 = _____ + _____ + _____

③ 304 = _____ + _____

④ 520 = _____ + _____

Fill in the missing values using the partial-sums strategy.

⑤
```
    243
  + 151
  
  _____ (200 + 100)
  
  _____ (40 + 50)
  
 +_____ (3 + 1)
```

⑥
```
    126
  + 315
  
  _____ (100 + 300)
  
  _____ (20 + 10)
  
 +_____ (6 + 5)
```

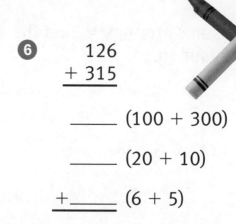

Practice
Fill in the missing values using the partial-sums strategy.

7. 436
     + 261

___ (___ + ___)
___ (___ + ___)
+ ___ (___ + ___)

8. 753
     + 138

___ (___ + ___)
___ (___ + ___)
+ ___ (___ + ___)

Find each sum using the partial-sums strategy.

9. 642
     + 153

10. 320
     + 246

11. 343
     + 111

12. 227
     + 516

Reflect
Use the partial-sums strategy to find the sum. Explain each of your steps.

   743
+ 352

_____

_____

_____

Using Partial Sums • Lesson 4

Using Partial Sums

Lesson 5 Review

This week you explored more addition strategies. You used partial sums and Base-Ten Blocks with renaming to find sums.

Lesson 1 Use the partial-sums strategy to find each sum.

①   38
    + 55

    ____ (30 + 50)
  +____ (8 + 5)

②   138
    + 312

    ____ (100 + 300)
    ____ (30 + 10)
  +____ (8 + 2)

Lesson 2 Use the partial-sums strategy to find each sum.

③   37
    + 44

    ____ (____ + ____)
  +____ (____ + ____)

④   536
    + 368

    ____ (____ + ____)
    ____ (____ + ____)
  +____ (____ + ____)

Reflect
Name the partial sums for the problem below.
Find the sum.

    556
  + 245

40 Addition • Week 4

Lesson 3 Use Base-Ten Blocks to create a model for each number. Rename ones to tens and tens to hundreds as needed. Draw your model. Find the sum.

⑤   64
    + 27

⑥   34
    + 19

⑦   78
    + 17

Lesson 4 ## Practice
Use Base-Ten Blocks to create a model for each number. Rename ones to tens and tens to hundreds as needed. Find the sum.

⑧   177
    + 286

⑨   454
    + 277

⑩   194
    + 268

Reflect
Which problem below does not require any renaming to find the sum? Explain.

⑪   31
    + 55

⑫   268
    + 513

⑬   344
    + 292

Using Partial Sums • Lesson 5 Review

Week 1: Using Doubles and Near Doubles

Practice

Use doubles facts to find the sums of each near-doubles fact.

1 If 20 + 20 = _____, then

20 + 21 = _____.

2 If 55 + 55 = _____, then

55 + 56 = _____.

Use doubles facts to find the missing addend.

3 12 + _____ = 23

4 14 + _____ = 29

Use the Number Construction Mat and Base-Ten Blocks to create a model of the addition problems. Find each sum.

5 14 + 10 = _____

6 46 + 10 = _____

Using a 100 Chart, find each sum.

7 47 + 9 = _____

8 19 + 9 = _____

Use +10 combinations to help you decide if the missing addend is 9 or 11. Circle your choice.

9 18 + (9 or 11) = 29

10 22 + (9 or 11) = 31

Using doubles facts, near-doubles facts, +10 combinations, +9 combinations, or +11 combinations, make an addition problem for the following sum. Write the strategy used.

11 39 = _____ + _____   _____

42 **Addition** • Week 1 Practice

Week 2 Using Nice Numbers

Practice

Circle the numbers below that are considered "nice numbers."

1) 18   40   96   11   29   70   35   10   24   33   64

Write a new addition sentence by sharing the values. Be sure that a "nice number" is in the new sentence.

2) 56 + 41

New addition sentence

_____

3) 17 + 62

New addition sentence

_____

Use a Number Construction Mat and Base-Ten Blocks or a Place Value Mat to build each addition problem. Say how many tens and ones are modeled.

4) 23 + 74

_____ tens and _____ ones = _____

5) 46 + 12

_____ tens and _____ ones = _____

6) 51 + 18

_____ tens and _____ ones = _____

7) 36 + 41

_____ tens and _____ ones = _____

Regroup each number into tens and ones. Then find each sum.

8) 42 = 40 + 2
     + 16 = 10 + 6
     _____

9) 26 = _____
     + 61 = _____
     _____

Addition • Week 2 Practice   **43**

Week 3 — Using Grouping

Practice

Answer the following questions.

1. Does 6 + 4 = 4 + 6? _____   If yes, what is the sum? _____

2. What do you notice about the problem above?

   _____

Reorder the numbers in each addition problem to make "nice numbers." Then find the sum.

3. 3 + 6 + 7 = _____

4. 9 + 8 + 2 + 1 = _____

Reorder the numbers in the addition problem to make "nice numbers." Then find the sum.

5. 7 + 5 + 13 + 4 + 5

   (_____ + _____) + (_____ + _____) + _____

   _____ + _____ + _____ = _____

Solve the word problem using addition. Write the number sentence.

6. Travis sets aside time for reading every night. Four nights ago, Travis read 26 pages of a book. He read 31 pages three nights ago and 21 pages two nights ago. Finally he read 30 pages last night. How many pages did Travis read over the last four nights?

   _____

44 Addition • Week 3 Practice

Using Partial Sums

Practice

Week 4

Use the partial-sums strategy to find each sum.

1. 284
     + 685

   _____ (200 + 600)

   _____ (80 + 80)

   +_____ (4 + 5)
   ─────

2. 158
     + 421

   500 (____ + ____)

    70 (____ + ____)

   + 9 (____ + ____)
   ─────

Use the partial-sums strategy to find each sum.

3. 46
     + 13

   _____ (____ + ____)

   +_____ (____ + ____)
   ─────

4. 824
     + 115

   _____ (____ + ____)

   _____ (____ + ____)

   +_____ (____ + ____)
   ─────

Use Base-Ten Blocks to create a model for each number. Rename ones to tens and tens to hundreds as needed. Draw your model. Find the sum.

5. 26
     + 37

6. 14
     + 38

7. 629
     + 187

8. 557
     + 385

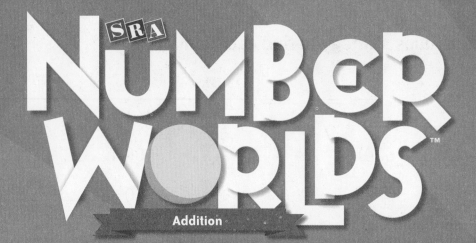

Unit 3 Workbook

SRAonline.com

Level E

R5313X.01